BYS 300 - Cell and Development Biology

Concepts and Investigations

Dr. Maria Ragland Davis

Edited by Jonathan Sullivan, Erika Knott, and Leland J. Cseke*

* Corresponding Author
Leland J. Cseke, Ph.D.
The University of Alabama in Huntsville
Department of Biological Sciences
Huntsville, AL 35899

Australia • Brazil • Japan • Korea • Mexico • Singapore • Spain • United Kingdom • United States

BYS 300 - Cell and Development Biology: Concepts and Investigations

Dr. Maria Ragland Davis

Senior Manager, Student Engagement:
Linda deStefano

Manager, Student Engagement:
Julie Dierig

Marketing Manager:
Rachael Kloos

Manager, Premedia:
Kim Fry

Manager, Intellectual Property Project Manager:
Brian Methe

Senior Manager, Production:
Donna M. Brown

Manager, Production:
Terri Daley

© 2015, 2011 Dr. Maria Ragland Davis

ALL RIGHTS RESERVED. No part of this work covered by the copyright herein may be reproduced, transmitted, stored or used in any form or by any means graphic, electronic, or mechanical, including but not limited to photocopying, recording, scanning, digitizing, taping, Web distribution, information networks, or information storage and retrieval systems, except as permitted under Section 107 or 108 of the 1976 United States Copyright Act, without the prior written permission of

Dr. Maria Ragland Davis

> For product information and technology assistance, contact us at
> **Cengage Learning Customer & Sales Support, 1-800-354-9706**
> For permission to use material from this text or product, submit all requests online at **cengage.com/permissions**
> Further permissions questions can be emailed to
> **permissionrequest@cengage.com**

ISBN: 978-1-305-28332-9

WCN: 01-100-101

Cengage Learning
20 Channel Center Street
Boston, MA 02210
USA

Cengage Learning is a leading provider of customized learning solutions with office locations around the globe, including Singapore, the United Kingdom, Australia, Mexico, Brazil, and Japan. Locate your local office at: **www.international.cengage.com/region.**

Cengage Learning products are represented in Canada by Nelson Education, Ltd.

For your lifelong learning solutions, visit **www.cengage.com/custom.**

Visit our corporate website at **www.cengage.com.**

In Loving Memory

Of

Dr. Maria Ragland Davis

Table of Contents
BYS 300 - Cell and Developmental Biology
Concepts and Investigations

Writing a Good Laboratory Report	1
Lab 1: Microscopy Analysis of Cells	6
Lab 2: Cell Structure and Function	19
Lab 3: Diffusion, Osmosis, and the Functional Significance of Biological Membranes	30
Lab 4: Cellular Respiration	40
Lab 5: Confocal Microscopy	49
Lab 6: Photosynthesis Capturing Energy	51
Lab 7: Quantification of Cellular Protein Content, Part I	57
Lab 8: Quantification of Cellular Protein Content, Part II	62
Lab 9: Exercise: Kingdom Fungi	65
Lab 10: Development in Animals, Part I: Sea Urchin Development	70
Lab 11: Development in Animals, Part II: Chick Development	76

Writing a Good Laboratory Report

Dr. Maria R. Davis
Department of Biological Sciences
The University of Alabama in Huntsville

All students will be asked to submit a typed lab report for many of the labs to be performed in this course. In the reports, the students will summarize the objectives of the experiments performed, provide background information, describe the experimental procedures, and discuss their results. Writing these reports will allow the students to think critically about the experimental procedures, results and their interpretation; and perform literature searches for additional information in textbooks, databases, research or review articles, etc.

It is important to note that writing provides an extremely efficient framework for learning. In each report, the students will demonstrate their level of understanding of the experimental procedures and the involved concepts. The following are guidelines for writing a lab report. It is crucial that all students follow the guidelines provided and take advantage of the opportunity to
improve their scientific writing. All reports will be thoroughly evaluated by the laboratory instructors, who will also provide suggestions and comments.

Unless otherwise specified by the lab instructor, every lab report must include the following:

1. A title and your name
2. Abstract
3. Introduction
4. Materials and methods
5. Results
6. Discussion
7. Literature cited

1. **A TITLE AND YOUR NAME**   On separate cover sheet, write:

Your original title for the laboratory experiment
Your full name
BYS300L – Section number
Lab Instructor's name
Date

From this point on, the report should be written in third person and past tense.

2. ABSTRACT
This section is an overview of the entire report. Readers should be able to read this section and know what your experiment was, why it was important, and what the results were. There should be approximately two sentences about each of the other sections here ("Approximately," because some people are better at properly condensing information than others).

3. INTRODUCTION
This should be the second largest section of your paper.

This section is the introduction to the experiment. You should tell the reader what parts of the cell are being studied and how, and provide background information about the organism/organelle. Find the important cell-related reason for performing this experiment. For example, if a substance is being extracted using a few methods and purified from a particular tissue, you could say that the ability to efficiently extract/purify the substance tells a lot about which method works best disrupting a cell membrane. Then give more information about the cell membrane and how it is "held together" and what types of treatments can disrupt it. In summary, first write about the important cell-biological implications of the experiment and give the appropriate background information, then think about the possible broader impacts if there
are any. Always try to conclude with a brief statement of what you are hoping to find out from your own experimental work and how your work fits into the "big picture," in order to tie this section to the rest of your report.

You should not tell the reader that you are doing an experiment in order to learn how to make graphs or tables, or learn how to do specific calculations (even if your lab manual says that is what you did).

Do not tell the reader about your results in this section. You are only introducing the concepts behind the experiment in this section. There will be plenty of opportunity for you to explain the data in the discussion section.

4. MATERIALS AND METHODS
This section is where you tell the reader what you did, exactly the way you did it. If we used a specific technique or equipment, such as the vortex mixer, then say "the tube was vortexed," not that "it was shaken."

As indicated above, the report should be written in third person and past tense. And, do not include lists. Instead, describe the procedures in a succinct and logical manner. Following is an example of the way this section should be written:

Tissue Preparation (for reference only, please do not use this example in your report!)
Large beetroots were selected and washed with tap water, and then with distilled water. Using a core borer, the instructor cut cylinders of 1 cm diameter for the entire class. 27 fresh and 3 frozen cylinders of approximately 2 cm length were trimmed by each group using a single edged razor blade and rinsed under running tap water for 10-15 min to remove the betacyanin present on the exterior of cylinders. This excess betacyanin was present due to a physical cut through the vacuoles of the roots when making the cylinders, and rinsing it off prior to the experiment ensured that it was not counted as extracted betacyanin.

Headings and subheadings keep you organized and also help the reader to better follow your experimental process. Finally, a well-thought-out content and logical order can earn you better grades for your lab report.

Use actual numbers; do not spell them out. Although you probably were taught in English class at some point that spelling out numbers is the correct way to write them, in the scientific world, people are reading your papers for the numbers. Do not hide them by spelling them out in this case. It is always important to remember your audience when you write a paper. The same goes for quantitative words such as milliliter or centimeter. In the scientific world, these are written as ml or cm.

If you perform data analysis with a computer program or calculator with standard formulas, you should indicate how the analysis was performed in the "Data analysis section":

Data Analysis
Standard deviation and mean calculations were performed using standard formulas in Microsoft Excel (or a TI-89 calculator, etc.)

5. RESULTS
This section requires a lot of time in preparation, prior to your discussion section. The amount of thought that you dedicate to this section can either make or break your report.

A variety of tables and graphs are useful in various situations. Now is a great time for you to learn which situations warrant which type of table or graph. If you are unsure about what type of table or graph to use, ask your lab instructor for suggestions.

Above each table and below each graph or figure, there should be a sentence or two that describes the table, graph or figure in detail. The reader should be able to completely understand all axis, columns, rows, units; and what the graph, figure or table shows. Any important trend in the data should be clearly indicated in the title for that figure, graph or table. The results section reports experimental data and trends in the data with no personal interpretation of the significance of these trends.

6. DISCUSSION

This should be the largest section of your paper. It is here where you will "pull it all together" and tell the reader what your data mean: Not in the introduction, not in the results, here!

In this section, you should start with the detailed information from the results sections and end by bringing that information together with your introductory material. You should tell the reader how everything worked. What equipment did you use and how does that equipment work?
What type of data does it provide and how does that relate to the questions you are asking?
What does this information tell you about the part of the cell you are studying? Was your hypothesis correct?

7. LITERATURE CITED
- You must include AT LEAST three references.
- You can have NO references that come from emails or bulletin board questions/answers from the internet.
- You should cite your lab manual when appropriate

Some Literature Research Sources

Texts, Reviews
Annual Review of Biochemistry
Annual Review of Plant Physiology and Plant Molecular Biology
Annual Review of Cell Biology
Methods in Enzymology Methods in
Cell Biology

Journals
Biochemical and Biophysical Research Communications
Biochemical Journal
Biochemistry
Biochimica et Biophysica Acta
Cell
Cell Biology International Reports
Cell Membranes
Structure and Function of Sciences of the U.S.A.
Plant Cell European Journal of Cell Biology
Experimental Cell Research
FEBS Letters
Federation Proceedings: Federation of American
        Societies for Experimental Biology
Journal of Biological Chemistry
Journal of Cell Biology
Journal of Cell Science
Journal of Cellular Physiology
Journal of Electron Microscopy

Experimental Zoology
Journal of Histochemistry and Cytochemistry
Journal of Microscopy
Journal of Plant Physiology
Trends in Cell Biology

Molecular and Cellular Biology
Molecular Biology of the Cell
Nature (any)
Plant Physiology
Plant Science Letters
Proceedings of the National Academy
Protoplasma Cell
Science Cytobios The
Tissue and Cell
Trends in Biochemical Sciences

Additional Journals at Salmon Library

Biochemistry and Cell Biology
Current Opinion in Cell Biology
Current Opinion in Genetics & Journal of Botany
        Experimental Development Journal of
Developmental Biology
Genes and Development
Journal of Cell Biology
Journal of Cellular Neurosciences
Current Topics in Developmental Biology

On-Line Searches for Journal Articles
PubMed: make sure to use the link posted in "online databases" from our library:
http://www.ncbi.nlm.nih.gov.elib.uah.edu/sites/entrez?db=pubmed
www.scholar.google.com
National Center for Biotechnology Information (http://www.ncbi.nlm.nih.gov/)
Highwire Press (http://highwire.stanford.edu/)

UAH Salmon Library links for: (requires UAH ID for access)
e-journals
Web of Science
Science Direct
(Also a search for UAH Biology Databases will include databases that list complete articles)

Lab 1:
Microscopy Analysis of Cells
Philip Shelp

OBJECTIVES
1. Demonstrate proper care and use of a compound light microscope
2. Compare magnification, resolving power, and contrast
3. Demonstrate the proper technique of preparing a wet mount and using the compound microscope as an instrument of measurement
4. Describe and demonstrate the similarities and differences between prokaryotic and eukaryotic cells
5. Identify the parts of plant and animal cells and state the function of each

INTRODUCTON

In the 17th century Robert Hooke built a microscope powerful enough to see objects at greater magnifications than had previously been possible. While examining a thin piece of cork, he observed many individual units that reminded him of the small cubicles that monks lived in. He called these units cells. Other scientists examined different plants and animals and realized that all living things are composed of cells. This laboratory session will give you an opportunity to see examples of cells and help you appreciate why an understanding of this basic unit of life is so important. The unaided human eye can detect objects as small as 0.1 mm in diameter. Most cells are between 0.01 mm and 0.1 mm in diameter and cannot be seen without a microscope. The light microscope, by virtue of its lens system, can extend our vision a thousand times so that objects as small as 0.1 micrometers (μm) in diameter can be seen.

Microscopy involves three basic concepts:

> **Magnification**: The degree to which the image of a specimen is enlarged.
> **Resolving power**: How well specimen detail is preserved during the magnifying process.
> **Contrast**: The ability to see specimen detail against its background. Stains and dyes are added to sections of biological specimens to increase contrast.

There are two types of microscope slides: 1) the permanent slide and 2) the temporary wet mount. The wet mount is prepared in the laboratory using water as the suspension medium. The permanent slide is usually purchased from a biological supply house and is sealed with permanent glue.

The microscope is an expensive precision instrument. When removing the microscope from the storage area, always grasp it with both hands. Place one hand around the arm and the other hand firmly under the base. Hold it close to your body for stability. Once you reach your work area, set the microscope down gently on the table with the arm toward you (*Figure 1.1*).

1.1 Exercise: Label the parts of the Compound Microscope

A. Support Structures

Arm: Functions as the handle of the microscope.

Stage: Large platform just below the revolving nosepiece. The microscope slide is placed on the stage with the specimen positioned directly over the opening through which light may pass.

Stage clips: Attached to the stage and are used to secure the slide in position.

Base: Lowermost part of the microscope and is in contact with the table.

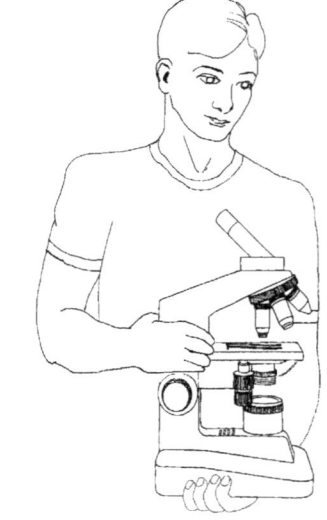

Figure 1.1 Carrying the compound microscope

B. Lighting

Light and light switch: Located in the base of the microscope.

Diaphragm: Shutter-like device below the stage that regulates the amount of light passing through a specimen. A lever regulates the size of the opening in the diaphragm.

Condenser lens: Located between the diaphragm and the stage aperture. This structure converges light rays from the light source so that they pass through the specimen on the slide and into the objective lens.

C. Focus

Coarse adjustment knob: Larger knob used to make large movements of the objective lens. Use this lens to locate your material.

Fine adjustment knob: Located either on top of the coarse adjustment knob or located separately. Manipulating this knob clarifies detail of your material.

D. Optics

The compound ocular microscope has at least two lens systems: an eyepiece that you look into and an objective that scans the specimen.

Eyepiece lens: Located in the upper end of the body tube and focuses light on the retina of the eye. The power of the eyepiece is usually 10X.

Oil immersion lens: May be attached to the revolving nosepiece. This lens will magnify 100X. A drop of immersion oil is used between the lens and the slide.
(We will not use this during this lab)

Objective lenses: Attached to the revolving nosepiece. The number and magnification of the objective lenses will vary with the type of microscope. The objective lenses are housed in one end of several steel tubes that are threaded into the revolving nosepiece. The desired objective lens is placed in position by rotating the nosepiece until it clicks into place. Most microscopes have three objective lenses: scanning lens (4X), low power lens (10X), and high power lens (40X).

Place the appropriate labels for each part of the microscope on *Figure 1.2*.

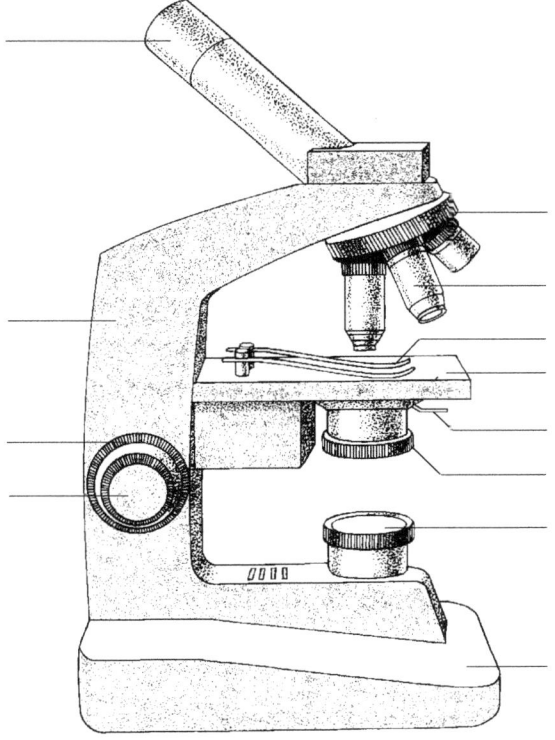

Figure 1.2 Compound light microscope

1.2 Exercise: Calculating Microscope Magnifications

Calculate the total magnification for each lens combination (*Table 1.1*).

Lens	Lens Power	Total Magnification
Unaided eye	1X	1X x 1X = 1X
Eyepiece	10X	1X x 10X = 10X
Scanning	4X	
Low Power	10X	
High Power	40X	
Oil Immersion	100X	

Hint: Remember that you must look through the eyepiece AND the objective lens for most set-ups.

Table 1.1 Magnification of each lens

1.3 Exercise: Focusing the Microscope

PROCEDURE
1. Rotate the coarse adjustment knob toward you. This will increase the working distance between the objective lens and stage.
2. Rotate the revolving nosepiece and click the scanning objective lens (4X) into position.
3. Place a slide on the stage of the microscope and secure it with the stage clips. Position the specimen directly over the center of the aperture of the stage.
4. View the microscope from the side and use the coarse adjustment knob to lower the objective lens. Keep the lens from coming in contact with the slide to prevent damage to the slide or lens.
5. Rotate the coarse adjustment knob toward you as you view the image through the eyepiece lens. As you view the specimen through the microscope, be sure to keep both eyes open. Then sharpen the image with the fine adjustment knob.
6. With the specimen in focus and positioned in the center of the field of view, rotate the nosepiece lens to low power (100X). Do not move the coarse adjustment knob. Only fine adjustment should be necessary to bring the specimen into sharp focus. The ability of the microscope to remain in focus when switching from one objective lens to the next highest power is called **parfocal**. It may be necessary to control the amount of light entering the objective lenses by adjusting the diaphragm. The higher the lens power the greater the amount of light is necessary. Adjust the diaphragm opening for optimum lighting.

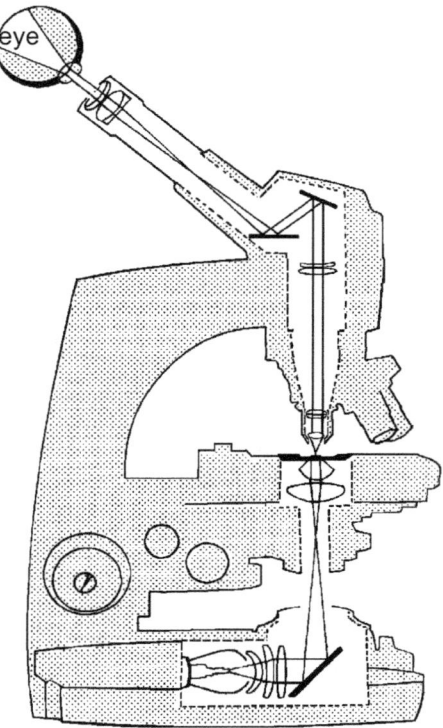

Figure 1.3 Focusing light on the retina of the eye

1.4 Exercise: Specimen Orientation

PROCEDURE
1. Place the permanent slide with the letter "e" right side up on the stage with the scanning power (40X) in place. The microscope slide label should be on your left side. Center the letter in the field of view and carefully bring it into focus. To focus the compound microscope, review *Exercise 1.3*.
2. Bring the "e" into focus under low power (100X).
3. Draw the "e" as you see it through the eyepiece lens on scanning power (40X) and low power (100X) (*Figure 1.4*).

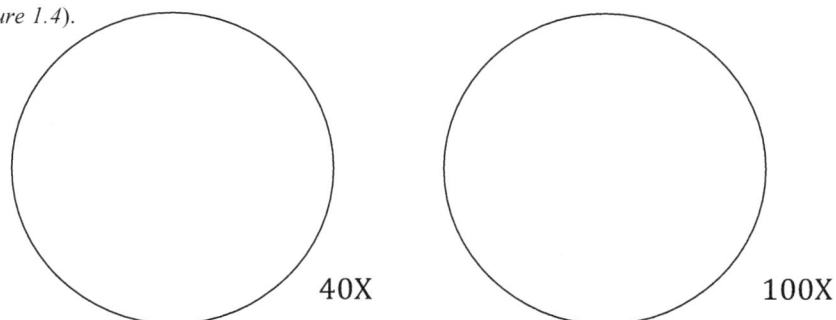

Figure 1.4 Letter "e" under scanning and low power

4. With either objective lens in place, move the prepared slide to the right while watching the image through the microscope.
5. Move the slide away from you in any direction.

POSTLAB QUESTIONS

Q1. In which direction does the image move when you move the stage to the left?

Q2. What is the relationship between the movement of the image and the object?

1.5 Exercise: Measure the diameter of the field of view

Data is most useful to science if it is quantitative (expressed in numbers). Thus, microscopic observations are more useful if the length and width of objects in the field of view can be measured. The field of view is the circle of light one sees in the microscope. Once the size of this field is determined, it is possible to estimate the size of any specimen by comparing it with the size of the field of view.

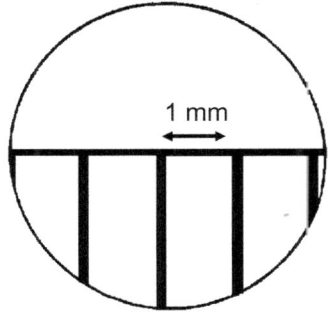

Figure 1.5 Diameter of the field of view of scanning power (40X)

PROCEDURE
1. Place a clean slide on the microscope stage and clip it in place.
2. Place a plastic ruler on top of the microscope slide and focus using the scanning power (40X).
3. Move the center of one-millimeter mark to the left side of the field of view as shown in *Figure 1.5* and measure the diameter of this field of view to the nearest 0.1 mm.
4. Because microscopic measurements are small, their dimensions are most useful when expressed as micrometers (μm). One millimeter is equal to 1000 micrometers (1 mm = 1000 μm).
5. It is difficult to use the metric ruler to measure the diameter of the field of view under low power (100X) because of its small diameter. Instead, use the equation below to calculate the diameter of the field of view. The diameter of the field of view is inversely proportional to the magnification of the lens. Let y equal the unknown diameter of the field of view under low power (100X), and d equal the known diameter of the field of view under scanning power.

$$\frac{40X}{100X} = \frac{y}{d}$$

The above equation can be used to determine the field of view of **low power (100X)** by solving for y as follows:
y = the unknown diameter in μm of the field of view
sp = magnification of the scanning power
d = diameter in μm of the scanning field of view m = magnification of the unknown field of view

The field of view of any microscope can be determined using the following equation:

$$y = \frac{sp \times d}{m} = \text{Example: } \frac{(40 \times 4500 \, \mu m)}{100X}$$

6. Calculate the size of the diameter of the field of view for each objective lens and record the data (*Table 1.2*). Remember that 1 mm = 1000 μm.

Lens	Total magnification	Diameter in mm	Diameter in μm
scanning			
low power			
high power			

Table 1.2 Diameter of the field of view for each objective lens

1.6 Exercise: Estimate the size of a specimen

Knowing the diameter of the field of view gives you a tool with which to estimate the size of any specimen you view under the microscope. If a specimen fills half the field of view, it is 1/2 of the diameter size. If it fills 1/4 of the field of view, it is 1/4 of the diameter size at that power.

PROCEDURE
1. Estimate the length of the specimen (*Figure 1.6*) under low power (100X).

2. Microscopic structures are often too small to be measured accurately under low power (100X). By knowing the diameter of the high power (400X) field of view, you can estimate the size of a specimen. Measure the length of the specimen (*Figure 1.6*) and answer the questions below.
    a. The diameter of the field of view under low power (100X): _____ µm
    b. The diameter of the field of view under high power (400X): _____ µm
    c. A specimen whose diameter is ¼ as wide as the field of view under low power (100X): _____ µm
    d. A specimen whose diameter is ¼ as wide as the field of view under high power (400X): _____ µm

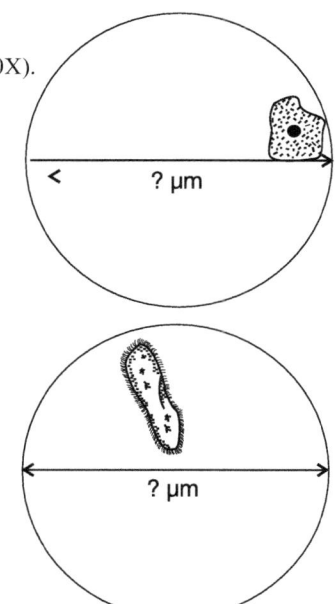

Figure 1.6 Specimen under low power (100X top) and high power (400X bottom)

1.7 Exercise: Determine depth of focus

The vertical distance that remains in focus at one time is called the **depth of focus**. Switch the microscope to high power and notice that the depth of focus is shallower with high power. As you focus up and down, you will get an idea of the specimen's three-dimensional form. The following exercise will enable you to reconstruct the three-dimensional structure of an object.

PROCEDURE
1. Obtain a slide with three colored threads mounted one over another.
2. Using low power (100X), find a point where all three threads cross (*Figure 1.7*). Then switch to high power (400X).
3. Slowly focus up and down to observe the depth of the threads.
4. Notice that when one thread is in focus the others seem blurred.
5. Determine the order of the colored threads and enter your results below.

Figure 1.7 Crossed threads demonstrating depth of focus

Top: **Middle:**   **Bottom:**

POSTEXCERCISE QUESTIONS

Q1. How did you determine the order of the threads?

Q2. How should you move the diaphragm in order to determine the order of colored threads the easiest?

Q3. Explain how depth of focus can be used to reconstruct structure.

Q4. Why would this technique be useful for examining cells?

CONCEPT: Cell Theory

With better microscopes, scientists were able to make more accurate observations about cells and gradually the **cell theory** developed. Currently, this theory has three parts:

1. All organisms are composed of one or more cells
2. The cell is the basic living unit of organization
3. All cells arise from pre-existing cells

Better microscopes enabled scientists to observe that while cells vary in organization, size, and function, they all contain the following structures:

1. A **plasma membrane** defining the boundary of the living material.
2. A region of **DNA (deoxyribonucleic acid),** which stores genetic information
3. A **cytoplasm** (everything inside the plasma membrane that is not part of the DNA region)

There are two basic types of cells: **eukaryotic**, those with a clearly defined nucleus and cell organelles, and **prokaryotic**, those without these structures. The Greek word karyon means kernel, referring to the nucleus. Thus, **prokaryotic** means "before a nucleus," while **eukaryotic** means "true nucleus" indicating the presence of a nucleus. Study *Table 1.3* to learn the differences between prokaryotic and eukaryotic cells.

Characteristics	Prokaryotic Cells	Eukaryotic Cells
Genetic material	Located in **nucleoid**, a region of cytoplasm not bounded by a membrane. Consists of a single molecule of DNA.	Located in a **nucleus**, a membrane-bound compartment within the cytoplasm. Made up of DNA molecules with proteins. Organized into chromosomes.
Cytoplasm	Small ribosomes. Photosynthetic membranes arising from the plasma membrane in some species.	Large ribosomes. Organelles, membrane-bounded compartments, which perform specific cell functions.

Table 1.3 Comparison of prokaryotic with eukaryotic cells

Prokaryotic cells, typical of modern bacteria and cyanobacteria, are believed to be similar to the first cells that developed billions of years ago. Eukaryotic cells have probably evolved from prokaryotes.

1.8 Exercise: Observing bacteria cells

PROCEDURE
1. Under the dissecting microscope, observe the culture plate containing bacteria growing on a nutrient medium. The dots you see are masses of cells called colonies, each originating from one cell that has divided many times.

2. Under the compound light microscope, observe the prepared slide of bacterial cells. The slide under this microscope is a preparation of stained individual bacterial cells.
3. Carefully draw what you see in the field of view (*Figure 1.8*). Record the magnification you use to view the bacteria.
4. Measure the approximate size (μm) of the bacterial cells.

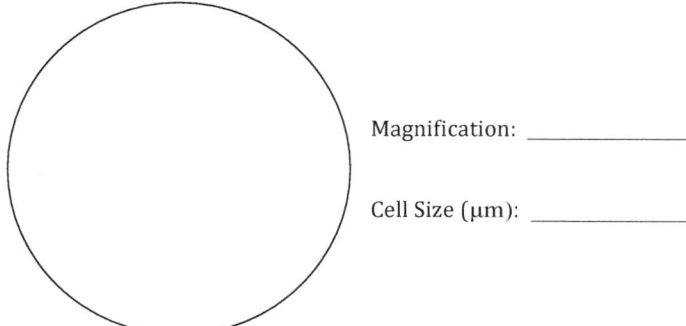

Magnification: _____

Cell Size (μm): _____

Figure 1.8 Drawing of prokaryotic cells

5. Examine the drawing of the bacterium (Figure *1.9*). The cell has a **cell wall,** a structure different from the wall of plant cells but serving the same primary function. The **plasma membrane** is flat against the cell wall and may be difficult to see. Look for two components in the **cytoplasm:** the small black dots called **ribosomes** give cytoplasm its granular appearance; the **nucleoid**, a relatively electron-transparent region (appears light) containing fine threads of DNA.
6. Using the terms shown in bold in step 5, label the bacterial structures (*Figure 1.9*).

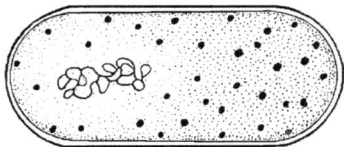

Figure 1.9 Representation of a bacterial cell

1.9 Exercise: Prepare a wet mount of human cheek cells

Animal cells are Eukaryotic.

PROCEDURE
1. Use a clean toothpick and gently scrape the inside of your cheek.
2. Stir the scrapings into a drop of water on a clean microscope slide and add a coverslip.
3. Methylene blue is a dye that will stain the cell's nucleus darker than the cytoplasm. Stain your sample by drawing a drop of stain under the coverslip by touching a piece of absorbent paper to the opposite side of the coverslip (*Figure 1.10*). Do not remove the coverslip.

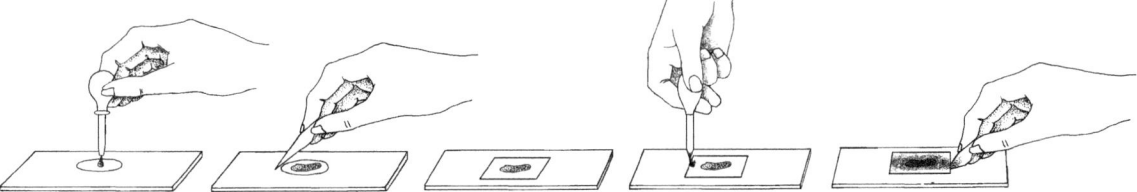

Figure 1.10 Preparing and staining a wet mount

4. Locate the cheek cells using low power, then switch to the high power for a more detailed view. Find the **nucleus**, a round centrally located body within each cell.

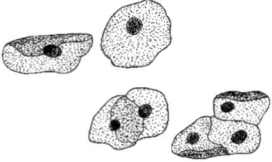

Figure 1.11 Human epithelium (cheek cells)

5. Carefully draw several cheek cells as they appear under the microscope (*Figure 1.12*). Label the cytoplasm, nucleus, and plasma membrane. Estimate the size of a typical cell. Record the cell size and magnification used.

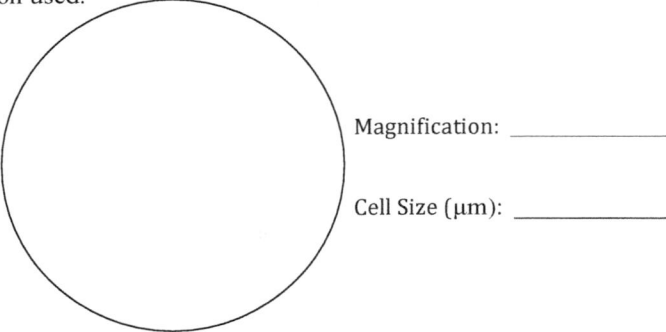

Magnification: _____

Cell Size (μm): _____

Figure 1.12 Drawing of the human cheek cell

1.10 Exercise: Prepare a wet mount of Elodea cells

The leaves of growing tips of Elodea will be used to study plant cell structure. Elodea leaves are only a few cell layers thick making it possible to view individual cells (*Figure 1.13*).
You will notice many spherical green **chloroplasts** in the cytoplasm. These organelles function in photosynthesis and are necessary for plant life. The **cell wall** is found only in plants and monera. It is a clear area outside of
the cytoplasm. The **plasma membrane** is not visible because it is pressed tightly against the cell wall and because it is beyond the resolving power of the light microscope.
The light source *may* heat the cells and cause **cytoplasmic streaming**. This is evident by the movement of chloroplasts along the cell wall. Microfilaments are responsible for this intracellular motion. Cytoplasmic streaming serves two functions: it positions the chloroplasts toward light and distributes heat throughout the cell.

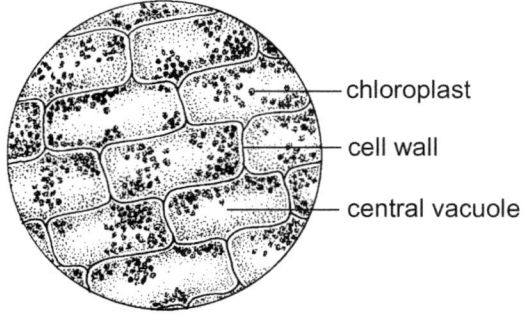

Figure 1. 13 Representation of a plant cell (Elodea)

Toward the middle of the cell, you will find the large, water-filled **central vacuole**. This structure may take up over half of the cell interior. **Any unequal distribution of chloroplasts you may see is due to the presence of a vacuole within the plant cell. Vacuoles frequently are large and occupy much of the cell volume, crowding the other contents of the cell out towards the cell wall.** The **nucleus**, within the cytoplasm, appears as a clear or slightly amber-colored body. It is slightly larger than the chloroplast.

PROCEDURE
1. Using forceps remove a young leaf from the growing tip of an Elodea plant and prepare a wet mount.
2. Examine the leaf structure with low power. Then study the detail of several cells using high power.
3. Carefully draw and label several Elodea cells in the field of view (*Figure 1.14*). Indicate where the plasma membrane is located in the cells.

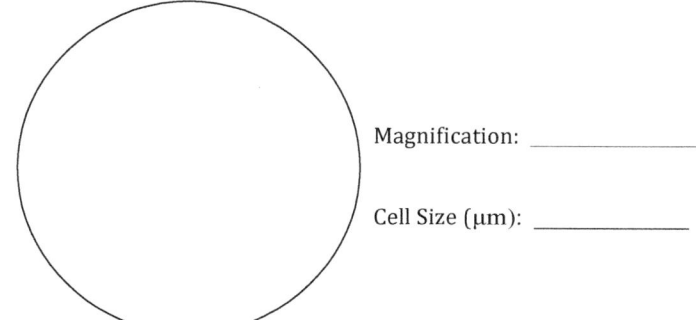

Magnification: _____

Cell Size (μm): _____

Figure 1.14 Drawing of Elodea cells

POSTEXCERCSE QUESTIONS

Q1. Describe the three-dimensional shape of the Elodea cell.

Q2. Describe the arrangement of the chloroplasts, nucleus and vacuole within the cell.

Q3. What is the structural importance of the cell wall?

1.11 Exercise: Onion bulb cells

PROCEDURE
1. Prepare a wet mount of onion epidermal cells using the technique described in *Figure 1.15*.
2. Stain the sample and then observe the wet mount with your microscope on low power.
3. Observe the specimen under high power.
4. Stain the specimen with iodine. The stain will increase the contrast and enable you to better view the nucleus, oil droplets, and cell wall.
5. The nucleus will be a large sphere within the cytoplasm. Examine the nucleus carefully and you will spot several nucleoli inside the nucleus. Nucleoli are the areas within the nucleus where rRNA (ribonucleic

acid) is being synthesized. The rest of the nucleus is largely DNA (deoxyribonucleic acid), the genetic material.

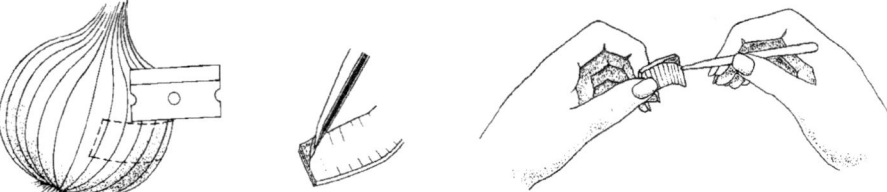

1. Using a razor blade, carefully cut a section from one bulb scale.

2. Using forceps, grasp the inner epidermis of the scale.

3. Remove this epidermis and prepare a wet mount of it.

Figure 1.15 Preparation of onion epidermis

6. Look for oil droplets in the form of granular material within the cytoplasm. The droplets are a form of stored food for the cell.
7. Carefully draw and label several onion epidermal cells (*Figure 1.16*). Draw the cell size to scale.

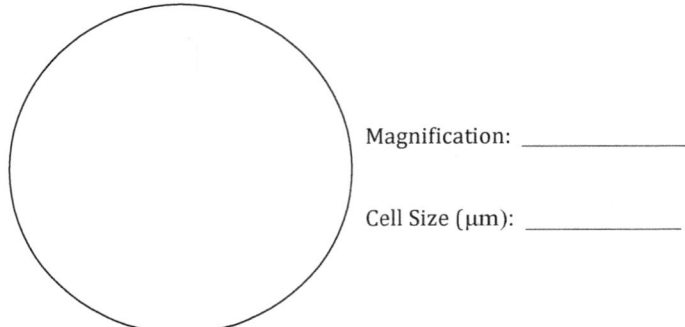

Magnification: _____

Cell Size (μm): _____

Figure 1.16 Drawing of onion epidermal cells

POSTEXCERCSE QUESTION

Q1. Based on the organelles you observed in both the onion and elodea, what process do you think occurs in the Elodea leaf cell but not the onion epidermal cell?

1.12 Exercise: Plant and animal cell organelles

The electron microscope has made it possible to observe great detail in cells. The following table contains a list of organelles that have been revealed through electron microscopy.

Structures found in both plant and animal cells	
Plasma membrane	Golgi body
Cytoplasm	Mitochondria
Nucleus:	Nucleolus
Nucleoplasm	Microtubules
Nuclear envelope	Ribosomes
Nuclear pores	Lysosome
DNA	Vacuole
	Vacuolar membrane
Endoplasmic Reticulum (ER): Rough ER (RER)	
Smooth ER (SER)	
Structures unique to plant cells	**Structures unique to animal cells**
Cell wall	Centriole pair
Large Central Vacuole	
Chloroplast	

Table 1.4 Comparison of plant and animal cell structures

Label the organelles in the figures below (*Figures 1.17 and 1.18*).

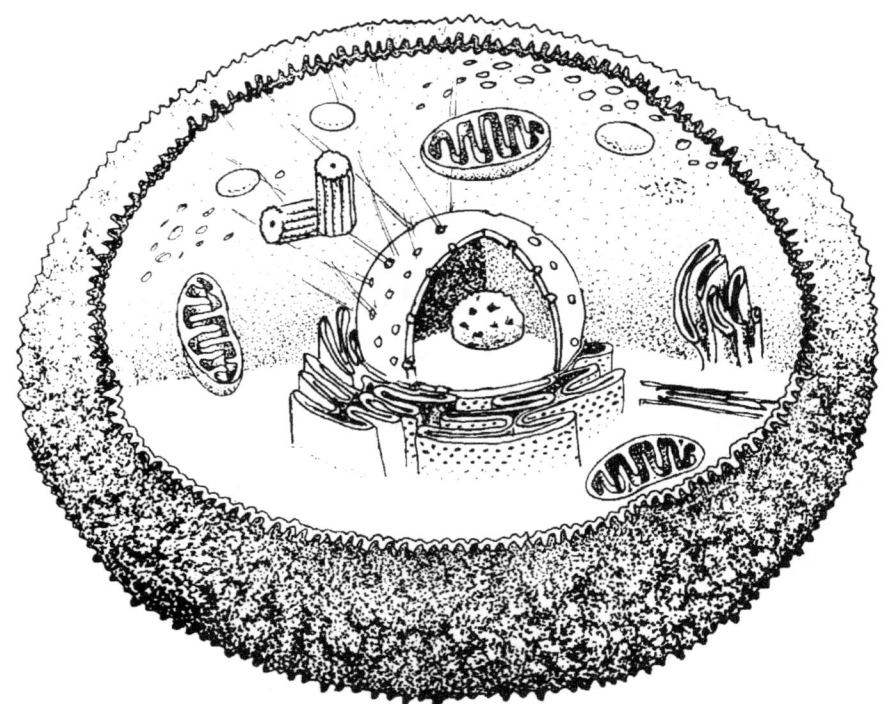

Figure 1.17 Typical animal cell

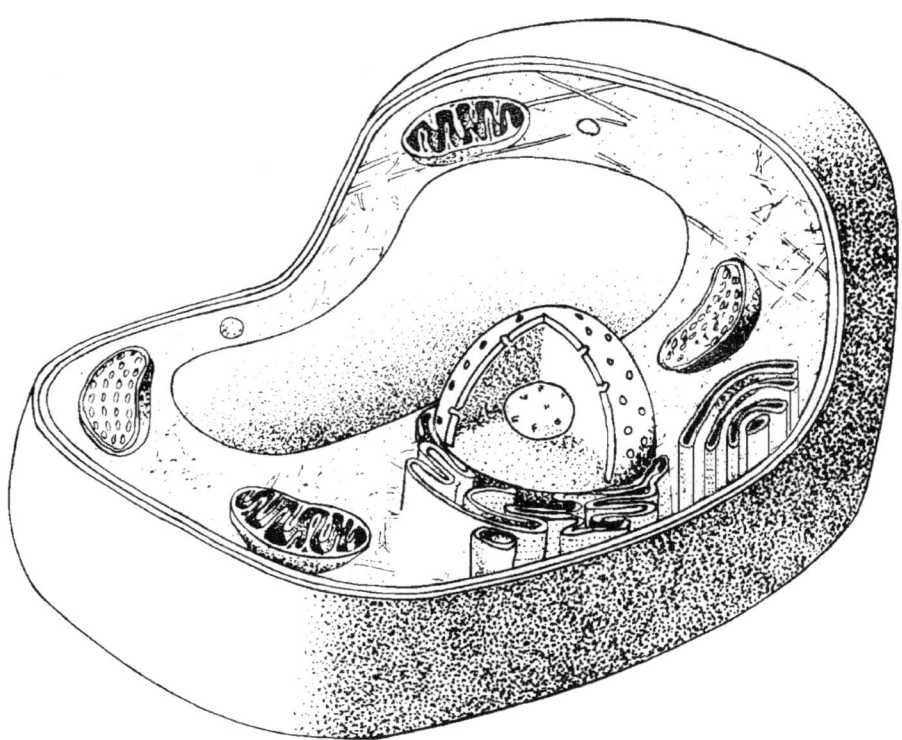

Figure 1.18 Typical plant cell

Lab 2:
Cell Structure and Function
Russell Skavaril, Mary Finnen & Steven Lawton

OBJECTIVES
1. State and describe the significance of the cell theory.
2. Describe the general characteristics of cells, including their size, range, and shape.
3. Contrast prokaryotic and eukaryotic cells. Draw and label a diagram of a prokaryotic cell and a eukaryotic cell.
4. Contrast plant and animal cells. Sketch a diagram of a plant cell and an animal cell.
5. Describe and list the functions of the principal cell organelles and structures:

   Nucleus   Rough endoplasmic reticulum
   Chloroplast   Smooth endoplasmic reticulum
   Nucleolus   Lysosome
   Vacuole   Cell wall
   Mitochondrion   Centriole
                              Ribosome

INTRODUCTON

The **cell theory** has three major components. The theory states that all organisms consist of at least one or more cells, cells are the smallest and the most fundamental living entity; and that a cell comes only from division of a previously existing cell. This theory also implies that cells are the biological structures within which the biological processes of metabolism occur and that cells contain hereditary material (nucleic acids).

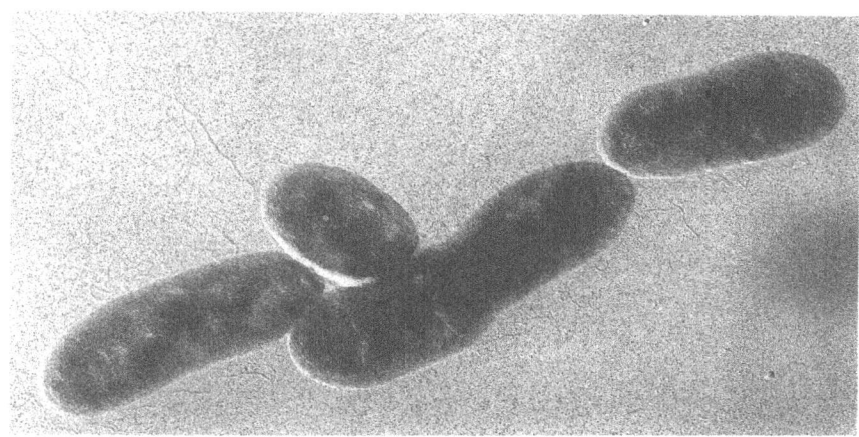

Figure 2.1 A shadowed electron micrograph preparation of Pseudomonas sp. showing flagella, 22,000X. (Electron microscopy photo by Dr. Robert A. Pfister.)

The cells that make up unicellular and multicellular organisms are highly varied. Cells come in a wide range of sizes — from the smallest, which is approximately 10 micrometers in diameter (2500 of these cells placed end to end would measure approximately 1 inch in length) to the largest cells, which are bird eggs. Different types of cells vary considerably in shape, from the round shape of normal red blood cells of humans to the long thread-like shape of human nerve cells. Nevertheless, all cells are classified as one of two basic types: prokaryotic and eukaryotic. **Prokaryotic cells** (eubacteria and archaebacteria) are simpler in structure than eukaryotic cells and are usually surrounded by a thick cell wall that provides protection and shape. Unlike eukaryotic cells, they lack a membrane-bound nucleus and organelles. DNA is present as a single circular molecule folded into a **nuclear area**. Where are most eukaryotic cells contain several compartmentalized areas, the plasma membrane is the only membrane found in most prokaryotes. Many bacteria are capable of movement by rotating long, thread-like **flagella** (*Figure 2.1*) that extend from the cell wall.

Eukaryotic cells (protists, fungi, plants, and animals) have membranes that surround an organized nucleus and various organelles with distinct functions. In general, cells of eukaryotic organisms are larger and more complex than prokaryotes.

Genetic Material (DNA) is contained within multiple linear **chromosomes** that carry genetic information and are contained within the **nucleus** (*Figure 2.2*) which segregates the chromosomes from the cytoplasm. The nucleus also holds the **nucleolus**, which produces granule-like ribosomes, massive proteins that are involved in protein synthesis. Prokaryotes also have ribosomes, but of smaller size. The nucleus is surrounded by a porous double membrane called the **nuclear envelope**, which controls the passage of materials into and out of the nucleus.

Figure 2.2 Freeze etching of Plasmodium gallinaceum, a blood parasite of the chicken, inside a chicken red blood cell. The electron micrograph, taken at a magnification of nearly 27,000X, shows, at upper left, the nucleus of the red blood cell and, at center, and the nucleus of the parasite. Both nuclei show nuclear pores. (Electron micrograph by Drs. Thomas Seed, Robert A. Pfister, and Julius P.

The outer nuclear membrane is continuous with the **endoplasmic reticulum** (ER), a system of flattened membranes throughout the cytoplasm. There are two types of ER. The first type, known as rough endoplasmic reticulum (RER), is associated with ribosomes and is involved with protein synthesis. The second type, known as smooth endoplasmic reticulum (SER), is not associated with ribosomes and is involved in lipid metabolism. The RER stores and segregates proteins prior to secretion from the cell. Before leaving the cell, proteins are frequently carried by vesicles from the RER to a Golgi complex, another membranous structure. Within the Golgi complex, proteins are modified and packaged prior to secretion or transportation to other organelles, such as lysosomes. Lysosomes are membrane-bound sacs containing enzymes that digest food molecules and worn-out cell components. The enzymes travel by vesicles from the Golgi complex and become deposited in the lysosome membrane.

Mitochondria are bound by an inner and outer membrane and function in eukaryotic cellular respiration wherein adenosine triphosphate (ATP) is produced. ATP leaves the mitochondria and provides energy for many cytoplasmic chemical reactions. Mitochondria have their own DNA, RNA, and ribosomes, which are involved in the production of new mitochondria.

Those structures described thus far (chromosome, nucleus, nucleolus, ribosome, nuclear envelope, endoplasmic reticulum, Golgi complex, lysosome, and mitochondrion) are found in essentially all plant and animal cells. Some organelles are characteristic of plants only. **Plastids** are similar to mitochondria in having a double membrane and their own nucleic acids and ribosomes, which are involved in self-reproduction. Plastids occur only in photosynthetic eukaryotes. Some plastids contain pigments that pro- duce bright colors in plant parts and other plastids store food. **Chloroplasts** are membrane-bound organelles that contain green chlorophyll. Chlorophyll is a special pigment that gives plants the capability of photosynthesis, a process that produces food (sugars) for the plant and, consequently, for any animal that eats the plant.

The **cell wall** is another structure absent in animal cells. It provides external support and protection for the plant cell. The wall, composed of cellulose and other fibers, is porous, allowing an exchange of materials between the plant cell and its environment. Plant cells also typically have a large **vacuole**, which functions in the storage of fluids and many dissolved sub- stances. A membrane that keeps stored materials separate from the cytoplasm, thereby increasing available surface area, surrounds the vacuole. Fluid within the vacuole exerts pressure against the cell wall, which helps support the bodies of softer plants.

Eukaryotic cells have a **cytoskeleton** that establishes cell shape and is involved in cell movement. The cytoskeleton is a network of protein filaments attached to the plasma membrane and various organelles. The thickest filaments of the cytoskeleton are the microtubules. **Microtubules** compose the centrioles that function in cell division. **Centrioles** are absent in higher plants. Short **cilia** and the longer flagella also consist of microtubules. They both function in cell movement or sweep liquid past a cell surface in immobile cells. Cilia and flagella do not occur in higher plants.

We shall examine some of these interesting and important structures.

2.1 Experiment: Cellular Diversity

MATERIALS

    Motor neurons, prepared slide

    Spirogyrum

PROCEDURE

In this exercise, you will examine the tremendous variation in the shape, size, and function of cells from different organisms.

    Motor neurons — animal cells shaped for specialized function.

    Spirostomum culture - the cell as an organism

1. Obtain a prepared slide of motor neurons. Place the slide on your microscope and examine it under both low and high power. Motor neurons are nerve cells that carry nerve impulses to muscles or glands resulting in some kind of action. The numerous cells on your slide probably came from the spinal cord of a mammal. The motor neurons are the larger; spider-like cells with a prominent nucleus and appendages that branch from the cell. Sketch a motor neuron in the space provided below (*Figure 2.3*).
2. Obtain a prepared slide of *Spirostomum*. This is a relatively large member of the group of unicellular organisms known as protozoa. *Spirostomum* moves by means of cilia, which are short, hair-like strands that cover the surface of the cell. The cilia beat rapidly in a regulated rhythm, propelling the cell through the water. Sketch a *Spirostomum* in the space provided below (*Figure 2.4*).

Figure 2.3 Drawing of a motor neuron   *Figure 2.4 Drawing of Spirogyrum*

PRELAB QUESTIONS

Q1. What are the three basic components of the cell theory?

1)

2)

3)

Q2. Compare the size, shape, and function of cells from one organism to another.

Q3. Define the major differences between eukaryotic and prokaryotic cells.

Q4. What organelle functions in eukaryotic cellular respiration (wherein ATP is produced)?

Q5. What is the difference between rough and smooth endoplasmic reticulum?

POSTLAB QUESTIONS

Q1. What do you think is the function of the appendages extending from the nerve cells?

Q2. What three structures did you view in the Elodea leaf cells, but not in the mammalian nerve cells?

Q3. Does Spirogyrum have a cell wall? How did you know?

Q4. How do nerve cells, Elodea leaf cells, and Spirogyrum compare in size?

2.2 Experiment: Prokaryotic Cells

Prokaryotic cells, such as bacteria and cyanobacteria, may be simple in structure, but many species are capable of biochemical activities, which make them very important to both the ecosystem and to humans. An important ecological biochemical activity is the decomposition of dead plant and animal bodies by bacteria, which release trapped nutrients back into the ecosystem where they are again used by other organisms. An important biochemical activity of humans is the use of certain bacteria in the processing of cheese and yogurt.

Cyanobacteria are a group of prokaryotes capable of making their own food by photosynthesis. Being prokaryotes, the photosynthetic pigment is not contained within chloroplasts, but is embedded in extensions of the cell membrane.

PROCEDURE
1. Use the flat end of a toothpick to place a tiny dab of yogurt onto a clean slide. Add a drop of water and mix thoroughly with the yogurt. Place a coverslip over the mixture and place the slide on your microscope. Examine under both low and high power. Looking carefully among the small bits of yogurt, you should see numerous bacteria that resemble tiny, thin filaments. These bacteria are *Lactobacillus*. They convert milk into yogurt. Note the size of these bacteria and draw them in the space provided below (*Figure 2.5*).

Figure 2.5 Drawing of Lactobacillus

POSTLAB QUESTIONS

Q1. Why is the plasma membrane of the various cell types you have observed not visible with your microscope?

Q2. How do you think the presence of a cell wall affects an organism?

2.3 Experiment: Identification of Unknown Cells

The various cells that you observed in the previous labs have given you enough experience to view un-known cells and to then classify them as prokaryotic, eukaryotic plant, or eukaryotic animal cells.

MATERIALS
3 Numbered microscopes/slides
Displays slides: animal, plant, and prokaryotic cells

PROCEDURE

Carefully look at the three unknown slides on display and try to identify them as representatives of one of the three general categories.

The following clues should help you make a decision:
- Are chloroplasts, or photosynthetic pigments, present or absent? (If observing a prepared slide, do not mistake green staining for photosynthetic pigment.)
- Is a cell wall present or absent?
- Is the cell relatively large or small?
- Does the cell interior appear modified and complex (organelles present) or does it seem to be unmodified and simple?
- Does the cell have an organized nucleus?

Record your decisions below by checking the appropriate answers.

Slide #1   prokaryotic,   eukaryotic   (plant / animal)

Slide #2   prokaryotic,   eukaryotic   (plant / animal)

Slide #3   prokaryotic,   eukaryotic   (plant / animal)

2.4 Experiment: Exploration of Cell Diversity

Pond water or a mixed culture of microscopic organisms provides an excellent hunting ground for exploring a variety of cell types, shapes, and structures.

MATERIALS
Pond water or mixed protozoa culture
Slides and coverslips
Dropper

PROCEDURE
1. Use a dropper to draw up debris from the bottom of a container of pond water or other culture for microscopic observation. If green alga is present, be sure to obtain some in your sample. Place a few drops of the sample on a slide, add a coverslip, and observe under low power.
2. Scan the slide for movement or anything that catches your eye. Observe any object of interest under high power if it improves visibility.
3. In the space provided below, draw at least four different specimens you find on the slide (*Figure 2.6*).

Figure 2.6 Drawings of various specimens

POSTLAB QUESTIONS

Q1. Do most of the specimens on your slide appear to be prokaryotic or eukaryotic? What criteria do you use to make this distinction?

Q2. Did you find mostly unicellular or multicellular organisms? Is it easy to distinguish between them?

Q3. Are any of the specimens you drew photosynthetic? How can you tell?

2.5 Experiment: Cells and Organelle Function

Many of the organelles within eukaryotic cells are too small to be easily seen with your microscope even though they are abundant in most of the cells you saw today. Use the table below and the electron micrographs in Figures 2.7 to 2.11 to familiarize yourself with some of these major cell organelles. With the help of your text, complete *Table 2.1* on page 29 by supplying the location (nucleus vs. cytoplasm), function, and origin (plant, animal, both) of the listed organelles.

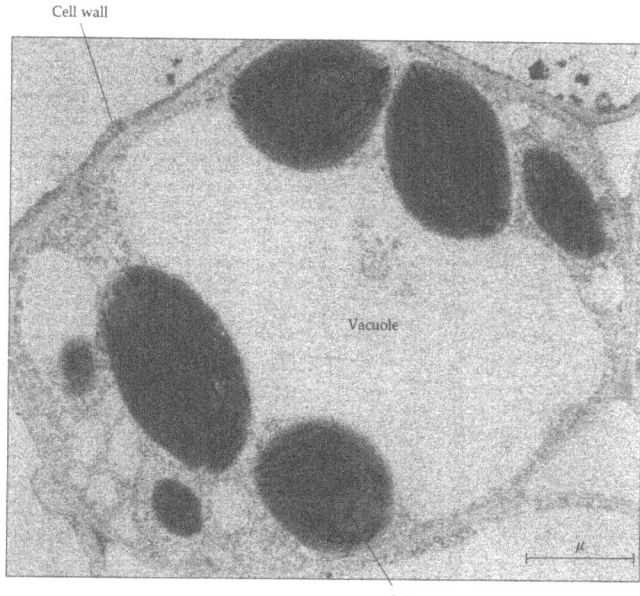

Figure 2.7 Electron micrograph of a grass cell showing the vacuole, cell wall, and chloroplasts typical of plant cells. (Biophoto Asso- ciates/Science Source/Photo Researchers)

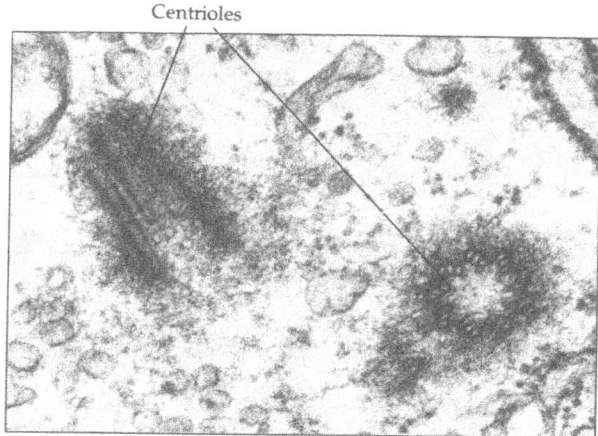

*Figure 2.8 Electron micrograph of two centrioles (approximately 80,000X), one sectioned lengthwise (left), the other in cross section (right).
(B.F. King, School of Medicine, Univ. of CA-Davis/BPS)*

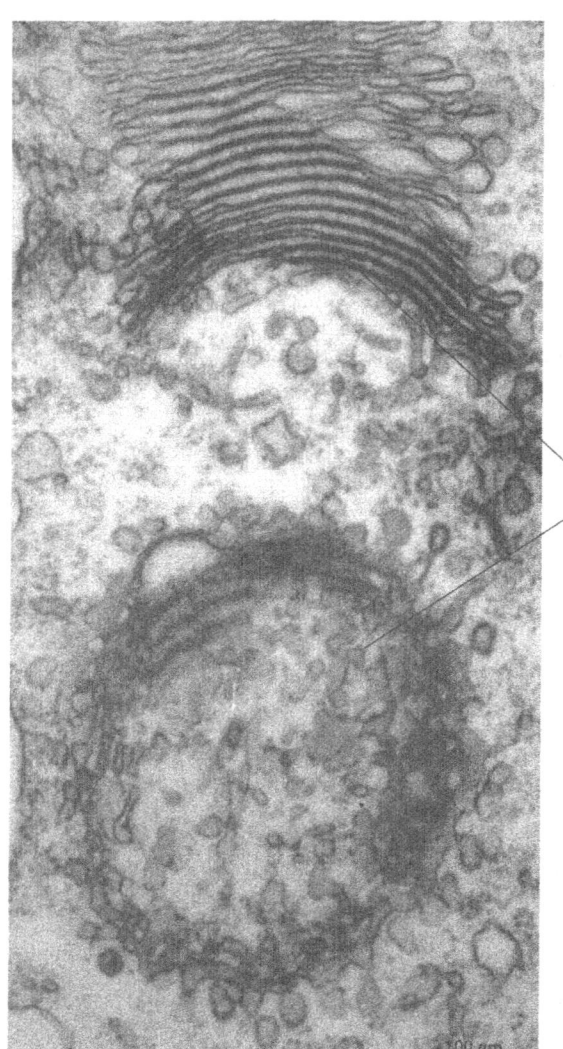

Figure 2.9 Electron micrograph of two Golgi complexes shown in cross section (top) and viewed from above (bottom).
(Biophoto Associates)

Figure 2.10 Electron micrograph of a nucleus (approximately 11,000X) and a darker nucleolus within its interior.
(Dr. Susumu Ito, Harvard Medical School)

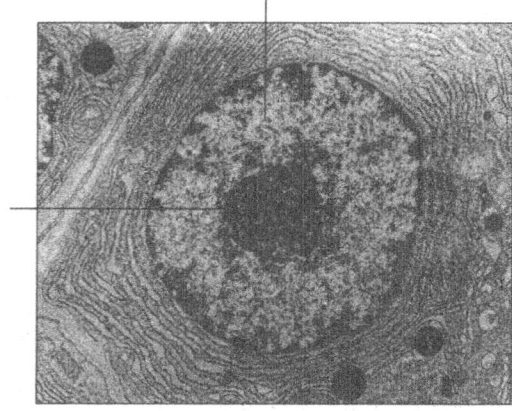

Figure 2.11 Electron micrograph of a lysosome digesting a starving cell's own mitochondria as a food source.
(Biophoto Associates)

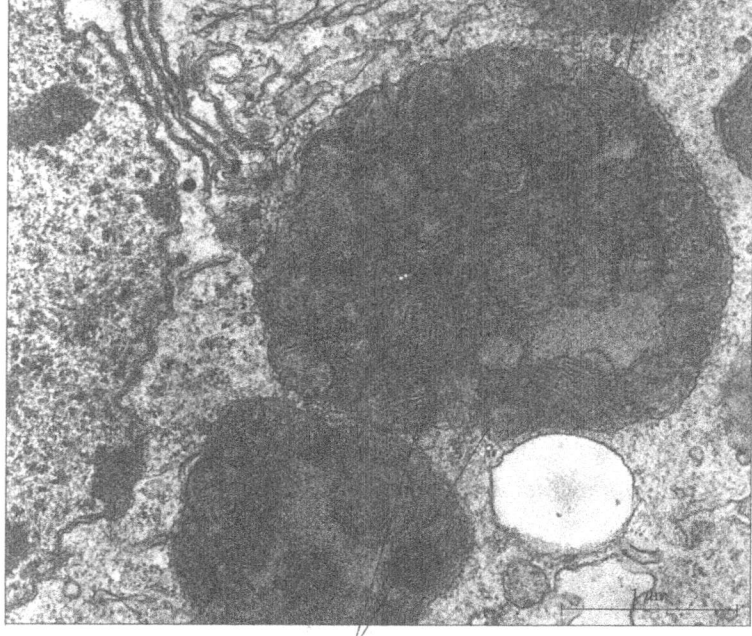

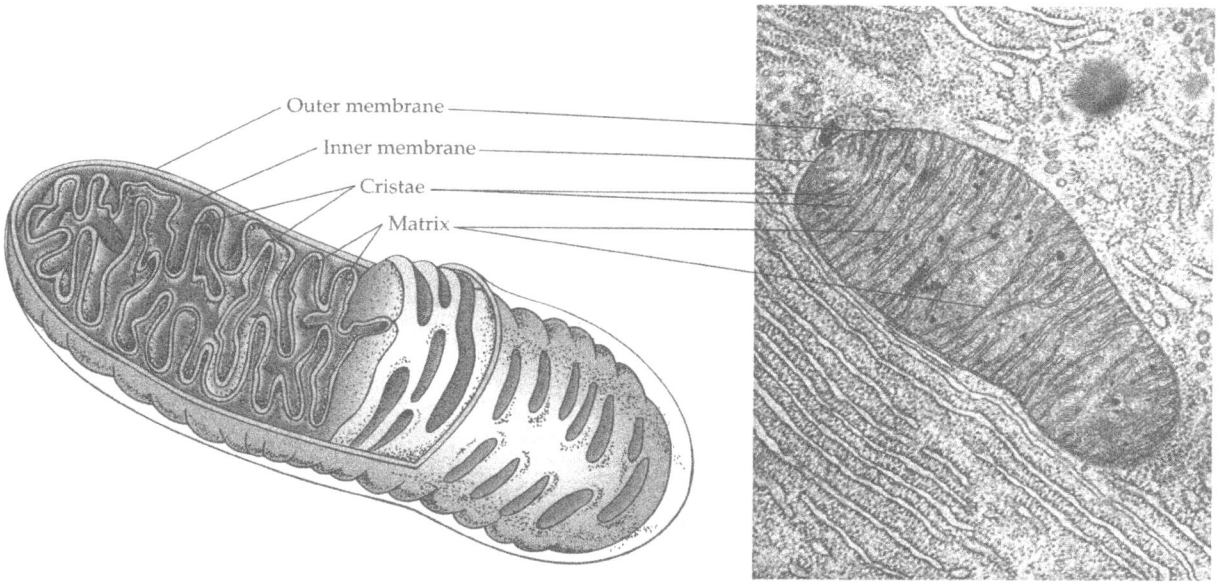

Figure 2.12 Drawing and electron micrograph of a mitochondrion (approximately 80,000X). (Keith Porter/ Photo Researchers, Inc.)

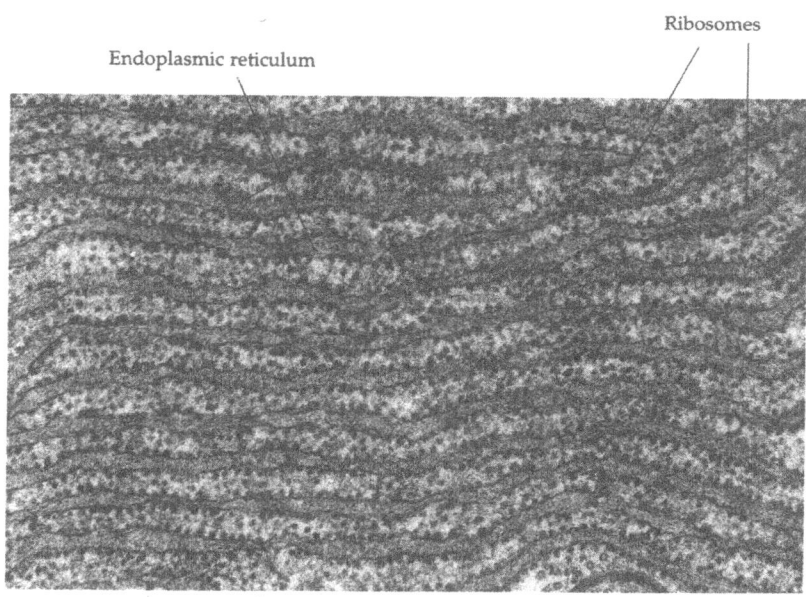

Figure 2.13 Electron micrograph of rough endoplasmic reticulum (approximately 70,000X). Note the many ribosomes associated with the endoplasmic reticulum membranes that give the "rough" appearance. Smooth endoplasmic reticulum appears similar but lacks the attached ribosomes.
(Courtesy of Dr. E. Anderson, Harvard Medical School)

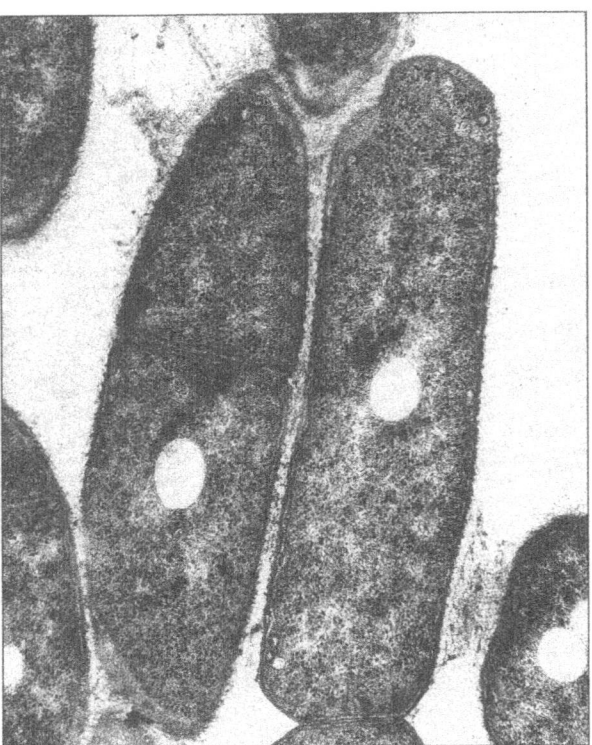

Figure 2.14 Thin section transmission electron micrograph of Bacillus subtilis, 30,000X, showing cell wall, cytoplasmic membrane, ribosomes, and polybetahydroxy butyric (PHB) acid granules. PHB is a cellular storage product and might possibly be important in the potential development of nonpetroleumbased plastic.
(Electron micrograph by Dr. Robert A. Pfister.)

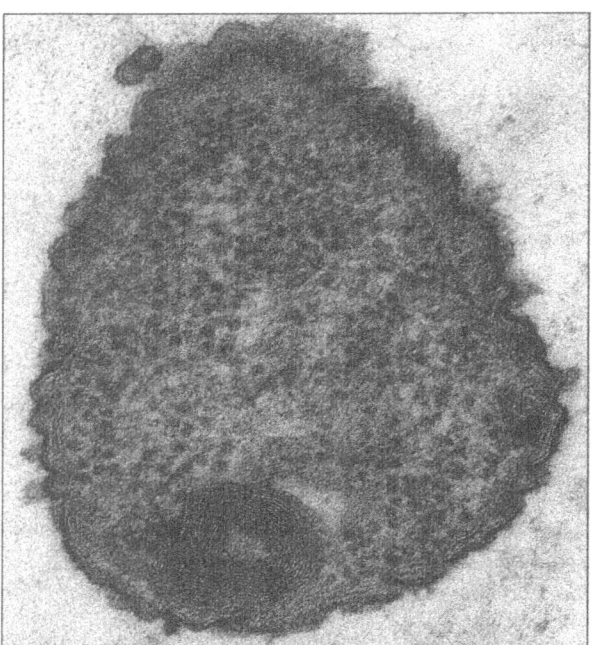

Figure 2.15 Thin section electron micrograph of Pseudomonas sp., 60,000X, showing mesosome. The mesosome is an infolding of the cytoplasmic membrane. The exact function of the mesosome is unknown.
(Electron micrograph by Dr. Robert A. Pfister.)

POSTLAB QUESTIONS

Q1. What is similar about the functions of RER, SER, and the Golgi complex? How might they interact?

Q2. Which organelles are exclusive to plants? Animals? How does this affect the functions of these organisms?

Table 2.1 Cell Structures

Structure	Location	Function	Origin
Cell wall			
Centriole			
Chloroplast			
Golgi complex			
Lysosome			
Mitochondrion			
Nucleolus			
Nucleus			
Ribosome			
Rough endoplasmic reticulum			
Smooth endoplasmic reticulum			
Vacuole			

Lab 3:
Diffusion, Osmosis, and the Functional Significance of Biological Membranes
James Perry, David Morton, and Joy Perry

OBJECTIVES
1. Define solvent, solute, solution, selectively permeable, diffusion, osmosis, concentration gradient, equilibrium, turgid, plasmolyzed, plasmolysis, turgor pressure, tonicity, hypertonic, isotonic, and hypotonic.
2. Describe the structure of cellular membranes.
3. Distinguish between diffusion and osmosis.
4. Determine the effects of concentration and temperature on diffusion.
5. Describe the effects of hypertonic, isotonic, and hypotonic solutions on Elodea leaf cells.

INTRODUCTION

Water is a great environment. Earthly life is believed to have originated in the water. Without it, life as we know it would cease to exist. Recently, the discovery of water in meteorites originating within our solar system has fueled speculation that life may not be unique to earth.

Living cells are made up of 75–85% water. Virtually all substances entering and leaving cells are dissolved in water, making it the solvent most important for life processes. The substances dissolved in water are called solutes and include such substances as salts and sugars. The combination of a solvent and dissolved solute is a solution. The cytoplasm of living cells contains numerous solutes, like sugars and salts, in solution.

All cells possess membranes composed of a phospholipid bilayer that contains different kinds of embedded and surface proteins. Look at *Figure 3.1* to get an idea of the complexity of a cellular membrane. Membranes are boundaries that solutes must cross to reach the cellular site where they will be utilized in the processes of life. These membranes regulate the passage of substances into and out of the cell. They are selectively permeable, allowing some substances to move easily while completely excluding others.

The simplest means by which solutes enter the cell is diffusion, the movement of solute molecules from a region of high concentration to one of lower concentration. Diffusion occurs without the expenditure of cellular energy. Once inside the cell, solutes move through the cytoplasm by diffusion, sometimes assisted by cytoplasmic streaming.

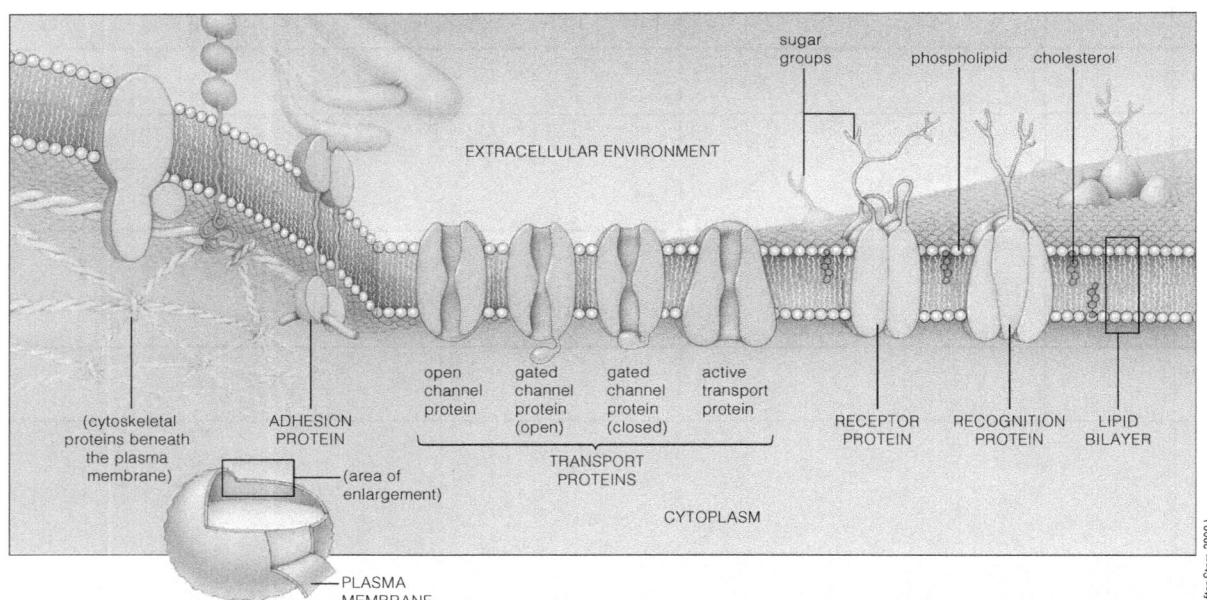

Figure 3.1 Artistic rendering of cutaway view of part of the plasma membrane.

Water (the solvent) also moves across the membrane. Osmosis is the movement of water across selectively permeable membranes. Think of osmosis as a special form of diffusion, one occurring from a region of higher water concentration to one of lower water concentration.

The difference in concentration of like molecules in two regions is called a concentration gradient. Diffusion and osmosis take place down concentration gradients. Over time, the concentration of solvent and solute molecules becomes equally distributed - the gradient ceases to exist.

At this point, the system is said to be at equilibrium; however, molecules are always in motion, even at equilibrium. Thus, solvent and solute molecules continue to move because of randomly colliding molecules. **BUT** It is important to note that at equilibrium there is no net change in their concentrations. This exercise introduces you to the principles of diffusion and osmosis.

> *Note: If Sections 3.2 and 3.3 are to be done during this lab period, start them before doing any other activity in this exercise.*

3.1 Experiment: Rate of Diffusion of Solutes

Solutes move within a cell's cytoplasm largely because of diffusion. However, the rate of diffusion (the distance diffused in a given amount of time) is affected by such factors as temperature and the size of the solute molecules. In this experiment, you will discover the effects of these two factors in gelatin (the substance of Jell-O®), a substance much like cytoplasm and used to simulate it in this experiment.

Two sets of three screw-cap test tubes have been half-filled with 5% gelatin; and 1 mL of a dye has been added to each test tube. Set 1 is in a 5°C refrigerator; set 2 is at room temperature. Record the time at which your instructor tells you the experiment was started:

MATERIALS
Per student:
Metric ruler
Per student group (table):

1 set of 3 screw-cap test tubes, in rack, each half-filled with 5% gelatin, to which the following dyes have been added: potassium dichromate, aniline blue, Janus green; labeled with each dye and marked "5°C"
1 set of 3 screw-cap test tubes, in rack, as above but marked "Room Temperature"

Per lab room: 5°C refrigerator

PROCEDURE
1. Remove set 1 from the refrigerator and compare the distance the dye has diffused in corresponding tubes of each set.
2. Invert and hold each tube vertically in front of a white sheet of paper. Use a metric ruler to measure how far each dye has diffused from the gelatin's surface. Record the distance that you measured in *Table 3.1*.
3. Determine the rate of diffusion for each dye by using the following formula:

   Rate of diffusion = distance traveled/elapsed time (hours)

   Experiment Start:
   Experiment End:
   Elapsed Time:

TABLE 3.1 Effect of Temperature on Diffusion Rates of Various Solutes				
	Set 1 (5°C)		Set 2 (Room Temp.)	
Solute (dye)	Distance (mm)	Rate	Distance (mm)	Rate
Potassium dichromate (MW = 294)[a]				
Janus green (MW = 511)				
Aniline blue (MW = 738)				

POSTLAB QUESTIONS

Q1. Which of the solutes diffused the slowest (regardless of temperature)?

Q2. Which diffused the fastest?

Q3. What effect did temperature have on the rate of diffusion?

3.2 Experiment: Osmosis

Osmosis occurs when a selectively permeable membrane separates different concentrations of water. One example of a selectively permeable membrane within a living cell is the plasma membrane. In this experiment, you will learn about osmosis using dialysis membrane, a selectively permeable cellulose sheet that permits the passage of water but obstructs passage of larger molecules. If you examined the membrane with a scanning electron microscope, you would see that it is porous; it thus prevents molecules larger than the pores from passing through the membrane.
Work in groups of four for this experiment.

MATERIALS

Per student group:
Four 15-cm lengths of dialysis tubing soaked in dH_2O
Eight 10-cm pieces of string or waxed dental floss
Dishpan half-filled with dH_2O
Paper toweling
Balance

25-mL graduated cylinder
4 small string tags
Marker
Four 400-mL beakers
Per lab room:
Source of dH_2O (at each sink)
15% and 30% sucrose solutions
Scissors (at each sink)

PROCEDURE
1. Obtain four sections of dialysis tubing, each 15 cm long that have been presoaked in dH_2O.
2. Fold over one end of each tube and tie it tightly with string or dental floss.
3. Attach a string tag or piece of tape to the floss at the tied end of each bag and number them 1–4. Slip the open end of the bag over the tip of a pipette and fill the four bags with the following solutions:

   Bag 1. 10 mL of dH_2O
   Bag 3. 10 mL of 30% sucrose Bag 2.
   10 mL of 15% sucrose Bag 4. 10 mL
   of dH_2O

4. As each bag is filled, force out excess air by squeezing the bottom end of the tube. When most air has been expunged, fold the end of the bag and tie it securely with another piece of string or dental floss.
5. Rinse each bag in the dishpan containing dH_2O; gently blot off the excess water with paper toweling.
6. Weigh each bag to the nearest 0.5g.
7. Record the weights in the column marked "0 min." in *Table 3.2*. Number four 400-mL beakers with a marker.

8. Place bags 1-4 in the correspondingly numbered beakers.

   Beakers #1-3
       Fill with 100mL of dH_2O
   Beaker #4
       Fill with 100mL of 30% sucrose solution.

9. After 15 minutes, remove each bag from its beaker, blot off the excess fluid, and weigh each bag.
10. Record the weight of each bag in *Table 3.2*. Return the bags to their respective beakers immediately after weighing.
11. Repeat Steps 11-12 for each bag at 30, 45, and 60 minutes.

At the end of the experiment, take the bags to the sink, cut them open, pour the contents down the drain, and discard the bags in the wastebasket. Pour the contents of the beakers down the drain and wash them according to the instructions given.

TABLE 3.2 Change in Weight as a Consequence of Osmosis

Bag	Bag Contents	Beaker Contents	Bag Weight (g)					Weight Change (g)
			0 min.	15 min.	30 min.	45 min.	60 min.	
1	dH_2O	dH_2O						
2	15% sucrose	dH_2O						
3	30% sucrose	dH_2O						
4	dH_2O	30% sucrose						

POSTLAB QUESTIONS

Q1. Make a qualitative statement about what you have observed.

Q2. Was the direction of net movement of water in bags 2–4 into or out of the bags?

Q3. Which bag gained the most weight? Why?

3.3 Experiment: Selective Permeability of Membranes

Dialysis tubing is a selectively permeable material that provides a means to demonstrate the movement of substances through cellular membranes.

MATERIALS
Per student group (4):
1 25-cm length of dialysis tubing, soaking in dH_2O

Two 10-cm pieces of string or waxed dental floss
Bottle of 1% soluble starch in 1% sodium sulfate (Na_2SO_4)

Dishpan half-filled with dH$_2$O
400-mL graduated beaker
Bottle of 1% albumin in 1% sodium chloride (NaCl)
8 test tubes
Test tube rack
Marker
25-mL graduated cylinder
Iodine (I$_2$KI) solution in dropping bottle

2% barium chloride (BaCl$_2$) in dropping bottle
2% silver nitrate (AgNO$_3$) in dropping bottle
Biuret reagent in dropping bottle
Scissors

Per lab room:
Series of 4 test tubes in test tube rack demonstrating positive tests for starch, sulfate ion, chloride ion, and protein

PROCEDURE

1. Obtain a 25-cm section of dialysis tubing that has been soaked in dH$_2$O.
2. Fold over one end of the tubing and tie it securely with string or dental floss to form a leak-proof bag.
3. Slip the open end of the bag over the tip of a pipette and fill the bag approximately half full with 25 mL of a solution of 1% soluble starch in 1% sodium sulfate (Na$_2$SO$_4$).
4. Remove the bag from the funnel; fold and tie the open end of the bag.
5. Rinse the tied bag in a dishpan partially filled with dH$_2$O.
6. Pour 100 mL of a solution of 1% albumin (a protein) in 1% sodium chloride (NaCl) into a 400-mL beaker.
7. Place the bag into the fluid in the beaker.
8. Record the time:
9. With a marker, label eight test tubes, numbering them 1–8.
10. Seventy-five minutes after the start of the experiment, pour 20 mL of the beaker contents into a clean 25-mL graduated cylinder.
11. Decant (pour out) 5 mL from the graduated cylinder into each of the first four test tubes.

    a) Perform the following tests, recording your results in *Table 3.3*. Your instructor will have a series of test tubes showing positive tests for starch, sulfate and chloride ions, and proteins. You should compare your results with the known positives. Test for starch. Add several drops of iodine solution (I$_2$KI) from the dropper bottle to test tube 1. If starch is present, the solution will turn blue-black.

    b) Test for sulfate ion. Add several drops of 2% barium chloride (BaCl$_2$) from the dropper bottle to test tube 2. If sulfate ions (SO-4) are present, a white precipitate of barium sulfate (BaSO$_4$) will form.

    c) Test for chloride ion. Add several drops of 2% silver nitrate (AgNO$_3$) from the dropper bottle to test tube 3. A milky-white precipitate of silver chloride (AgCl) indicates the presence of chloride ions (Cl^{2+}).

    d) Test for protein. Add several drops of Biuret reagent from the dropper bottle to test tube 4. If protein is present, the solution will change from blue to pinkish-violet. The more intense the violet hue, the greater the quantity of the protein.

12. Wash the graduated cylinder as instructed.
13. Thoroughly rinse the bag in the dishpan of dH$_2$O.
14. Using scissors cut the bag open and empty the contents into the 25-mL graduated cylinder.
15. Decant 5-mL samples into each of the four remaining test tubes.
16. Perform the tests for starch, sulfate ions, chloride ions, and protein on tubes 5–8, respectively.
17. Record the results of this series of tests in *Table 3.4*.
18. Discard contents of test tubes and beaker down sink drain. Wash glassware as instructed.

Discard dialysis tubing in wastebasket.

TABLE 3.3 Results of Tests for Substances in Beaker[a]		
	At Start of Experiment	After 75 min.
Starch	−	
Sulfate ion	−	
Chloride ion	+	
Albumin	+	

TABLE 3.4 Results of Tests for Substances in Dialysis Bag[a]		
	At Start of Experiment	After 75 min.
Starch	+	
Sulfate ion	+	
Chloride ion	−	
Albumin	−	

POSTLAB QUESTIONS

Q1. To which substances was the dialysis tubing permeable?

Q2. What physical property of the dialysis tubing might explain its differential permeability?

Q3. What do cells use to regulate transport across their membranes?

3.4 Experiment: Plasmolysis in Plant Cells

Plant cells are surrounded by a rigid cell wall, composed primarily of the glucose polymer, cellulose. Recall from Exercise 6 that many plant cells have a large central vacuole surrounded by the vacuolar membrane. The vacuolar membrane is selectively permeable. Normally, the solute concentration within the cell's central vacuole is greater than that of the external environment. Consequently, water moves into the cell, creating turgor pressure, which presses the cytoplasm against the cell wall. Such cells are said to be turgid. Many non-woody plants (like beans and peas) rely on turgor pressure to maintain their rigidity and erect stance.

In this experiment, you will discover the effect of external solute concentration on the structure of plant cells.

Tonicity describes one solution's solute concentration compared to that of another solution. The solution containing the lower concentration of solute molecules than another is hypotonic relative to the second solution. Solutions containing equal concentrations of solute are isotonic to each other, while one containing a greater concentration of solute relative to a second one is hypertonic.

MATERIALS

Per student:

Forceps
2 microscope slides
2 coverslips
Compound microscope

Per student group (table):

Elodea in tap water
Dropping bottle of dH_2O
Dropping bottle of 20% Sodium Chloride

PROCEDURE

1. With a forceps, remove two young leaves from the tip of an Elodea plant.
2. Mount one leaf in a drop of distilled water on a microscope slide and the other in 20% NaCl solution on a second microscope slide.
3. Place coverslips over both leaves.
4. Observe the leaf in distilled water with the compound microscope. Focus first with the medium-power objective and then switch to the high-dry objective.
5. Label the photomicrograph of turgid cells (*Figure 3.2*).
6. Now observe the leaf mounted in 20% NaCl solution. After several minutes, the cell will have lost water, causing it to become plasmolyzed. (This process is called plasmolysis.)

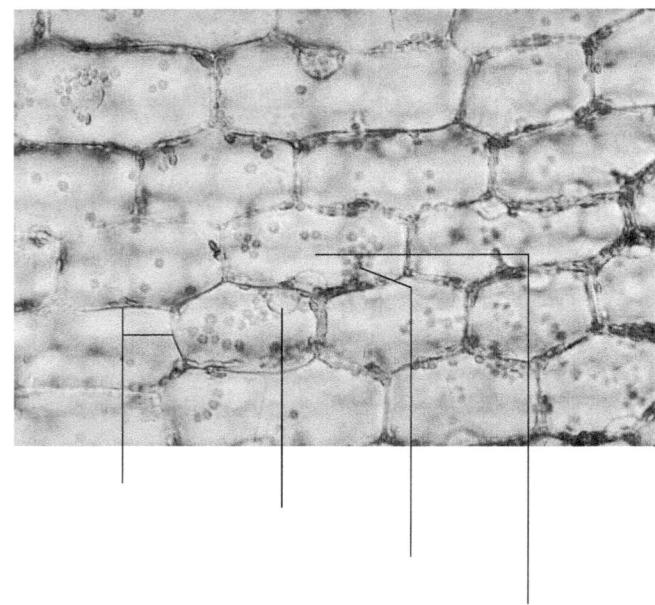

Figure 3.2 Turgid Elodea cells (400X) (Photo by J. W. Perry)

Label: Nucleus, Cell Wall, Chloroplast in Cytoplasm, Central Vacuole

POSTLAB QUESTIONS

Q1. If a selectively permeable membrane separates a hypotonic and a hypertonic solution, in which direction will the water move?

Q2. Name two selectively permeable membranes that are present within the Elodea cells and that were involved in the plasmolysis process.

1.

2.

3.5 Experiment: Determining the Concentration of Solutes in Cells

If you've done the previous experiments of this exercise, you now know that water flows into or out of cells in response to the concentration of solutes within the cells. But you might logically ask at this point how much solute is present in a typical cell. While the answer varies from cell to cell, a simple experiment enables you to determine the osmotic concentration in the cells of a potato tuber.

MATERIALS

Per student group (4):

Five 250mL beakers
Large Peeled Potato Tuber

Marker
Single-edge razor blades or paring knife
Metric Ruler

PROCEDURE

1. With the china marker, label the five 250-mL beakers with the concentrations of sucrose solution.
2. Pour about 100 mL of each solution into its respective beaker.
3. Peel the potato and then cut it into five 3-cm cubes (3 cm on each side).
4. Without delay, weigh each cube to the nearest 0.01 g. Record the weights in *Table 3.6*.
5. Place one cube in each beaker and allow it to remain there for a minimum of 30 minutes, longer if time is available.
6. After the experimental period has elapsed, remove each cube, one at a time, and blot it lightly but thoroughly with the paper toweling.
7. Weigh each cube and record its final weight in *Table 3.6*. Then calculate and record the weight loss or gain.
8. Calculate the percent change in weight by dividing the difference between the initial weight and final weight by the initial weight.

The cube with the lowest percentage of weight change is in a solution that most closely approximates the solute concentration of the cells within the potato tuber. Of course, most of the solute within the tuber is in the form of starch, and our experimental solution is sucrose. The results of this experiment indicate that the concentration of the solute, not the type of solute, is important for osmosis to occur.

POSTLAB QUESTIONS

Q1. Which concentration resulted in the greatest percentage change?

Q2. What was the approximate concentration of solute in the potato tuber?

Q3. Make a statement that relates the amount of water loss or gain to the concentration of the solute.

PRE-LAB QUESTIONS

_____ 1. If one were to identify the most important compound for sustenance of life, it would probably be
(a) salt
(b) $BaCl_2$
(c) water
(d) I_2KI

_____ 2. A solvent is
(a) the substance in which solutes are dissolved
(b) a salt or sugar
(c) one component of a biological membrane
(d) selectively permeable

_____ 3. Diffusion
(a) is a process requiring cellular energy
(b) is the movement of molecules from a region of higher concentration to one of lower concentration
(c) occurs only across selectively permeable membranes
(d) is none of the above

_____ 4. Cellular membranes
(a) consist of a phospholipid bilayer containing embedded proteins
(b) control the movement of substances into and out of cells
(c) are selectively permeable
(d) are all of the above

_____ 5. An example of a solute would be
(a) Janus green B
(b) water
(c) sucrose
(d) both a and c

_____ 6. Dialysis membrane is
(a) selectively permeable
(b) used in these experiments to simulate cellular membranes
(c) permeable to water but not to sucrose
(d) all of the above

_____ 7. Specifically, osmosis
(a) requires the expenditure of cellular energy
(b) is diffusion of water from one region to another
(c) is diffusion of water across a selectively permeable membrane
(d) is none of the above

_____ 8. Which of the following reagents does *not* fit with the substance being tested for?
(a) Biuret reagent   protein
(b) $BaCl_2$ starch
(c) $AgNO_3$ chloride ion
(d) albustix   protein

_____ 9. When the cytoplasm of a plant cell is pressed against the cell wall, the cell is said to be
(a) turgid
(b) plasmolyzed
(c) hemolyzed
(d) crenate

_____ 10. If one solution contains 10% NaCl and another contains 30% NaCl, the 30% solution is
(a) isotonic
(b) hypotonic
(c) hypertonic
(d) plasmolyzed, with respect to the 10% solution

3.1 Experiment: Rate of Diffusion of Solutes

1. You want to dissolve a solute in water. Without shaking or swirling the solution, what might you do to increase the rate at which the solute would go into solution? Relate your answer to your method's effect on the motion of the molecules.

3.2 Experiment: Osmosis

2. If a 10% sugar solution is separated from a 20% sugar solution by a selectively permeable membrane, in which direction will there be a net movement of water?

3. Based on your observations in this exercise, would you expect dialysis membrane to be permeable to sucrose? Why?

3.4 Experiment: Plasmolysis in Plant Cells

4. You are having a party and you plan to serve celery, but your celery has gone limp, and the stores are closed. What might you do to make the celery crisp (turgid) again?

5. Why don't plant cells undergo osmotic lysis?

6. This drawing (*Figure 3.3*) represents a plant cell that has been placed in a solution.
    a. What *process* is taking place in the direction of the arrows? What is happening at the cel- lular level when a wilted plant is watered and begins to recover from the wilt?

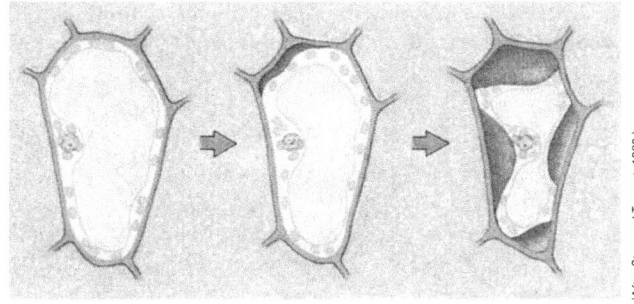

Figure 3.3 A plant cell that has been placed in a solution

Lab 4:
Cellular Respiration
Marni Fylling

OBJECTIVES
1. Understand the process that converts food energy into energy cells can use
2. Understand and identify reduction and oxidation reactions
3. Distinguish between aerobic and anaerobic reactions
4. Identify the products of alcoholic fermentation

INTRODUCTION

Plants do not need to eat. They obtain energy from the sun and use this light energy to produce chemical energy in a process called **photosynthesis**, the subject of the next lab. Organisms such as plants that synthesize their own organic molecules from inorganic material are **autotrophs** (auto = self, trophe = nourish). Like fungi and many other organisms, humans are **heterotrophs** (hetero = other). We must obtain our energy from other organisms by eating autotrophs, other heterotrophs, or both. All cells use ATP (adenosine triphosphate) to fuel their biological processes. Both heterotrophs and autotrophs obtain ATP by breaking down organic molecules in a process called **cellular respiration**.

Cellular respiration begins with small organic molecules, such as the sugar glucose. If you eat a piece of pizza, cellular respiration cannot begin until you have **digested** the pizza, or reduced it to smaller units. Proteins, polysaccharides, and lipids must be broken into smaller molecules that may enter a cell. Cellular respiration may then occur: the energy in the amino acids, monosaccharides, and fatty acids is converted into the chemical bonds of ATP.

During cellular respiration of glucose, the bonds between the carbon atoms in glucose are broken, and the energy stored in those bonds is used to form the high-energy bonds in ATP. In the process, the glucose molecule is broken down into carbon dioxide and water.

$$C_6H_{12}O_6 + 6\ O_2 \xrightarrow{\text{cellular respiration}} 6\ CO_2 + 6\ H_2O$$
$$\text{glucose + oxygen} \qquad\qquad\qquad \text{carbon dioxide + water}$$

The complete breakdown of glucose occurs in the four stages shown in *Figure 4.1*: **glycolysis**, the formation of **acetyl-CoA**, the **citric acid cycle**, and the **electron transport chain**. Glycolysis occurs in the cytosol of the cell; the rest of cellular respiration occurs in the mitochondria.

In this lab, you will perform experiments to identify products formed during two of the four stages of cellular respiration. Many different enzymes work to catalyze the numerous reactions that occur in cellular respiration. Two **coenzymes**, NAD$^+$ and FAD, also participate in these reactions. Unlike enzymes, coenzymes are not catalysts for chemical reactions; instead, they are arrier molecules. These coenzymestake electrons from glucose and eventually transfer those electrons to oxygen. **Oxidation** is the loss of an electron by a molecule, ion, or atom: a molecule, ion, or atom that loses an electron is oxidized. Reduction is the gain of an electron by a molecule, ion, or atom: one that accepts an electron is reduced. The electron transfer to oxygen in the electron transport chain is coupled with the formation of ATP. **Some of these experiments take time. Set up the carbon dioxide production experiment first, then the ethanol production experiment.**

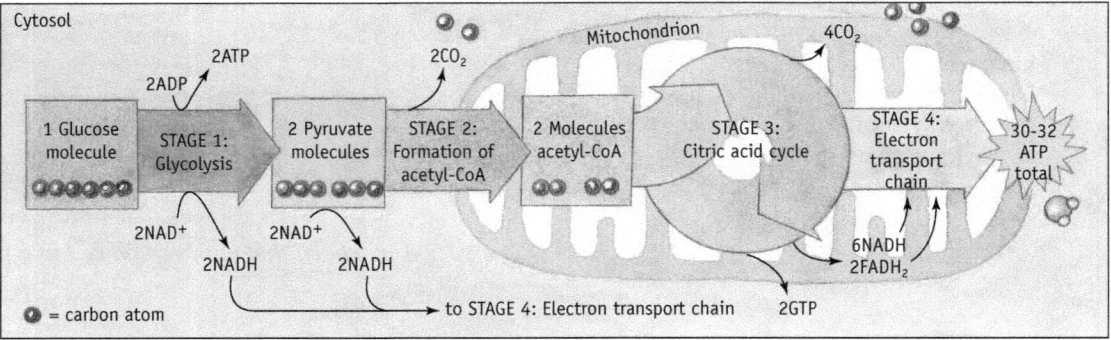

Figure 4.1 Overview of Cellular Respiration

Stages of Cellular Respiration

Glycolysis
The first stage in cellular respiration involves the breakdown of glucose, a six-carbon sugar, into two molecules of pyruvate, each with three carbon atoms. In this process, shown in *Figure 4.2*, some energy stored in the glucose is converted into two high- energy phosphate bonds in ATP. In addition, two molecules of the coenzyme NAD^+ are reduced: they each accept two electrons and a hydrogen ion to become two high- energy molecules of NADH.

In the presence of oxygen, NADH is oxidized back to NAD^+. It transfers its new electrons to the electron transport chain for the formation of ATP. Then, the NAD^+ helps break down another molecule of glucose into pyruvate, continuing the process of glycolysis.

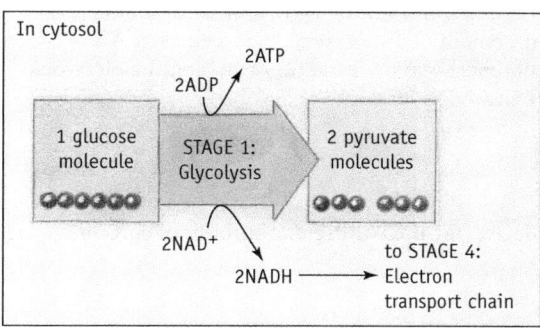

Figure 4.2 Glycolysis. Stage 1 of cellular respiration.

Fermentation
In the absence of oxygen, the cell still must oxidize its NADH, or glycolysis cannot continue. Have you ever exercised vigorously and experienced a tingling in your muscles? Your muscle cells needed more oxygen than your blood and lungs could deliver and had to oxidize NADH to continue glycolysis to get energy. In the absence of oxygen, NADH can donate its electrons to pyruvate, converting the pyruvate into lactic acid in a process called lactic acid fermentation. The accumulation of lactic acid in your muscles caused the tingling. As oxygen becomes available, the lactic acid converts back to pyruvate.

$$\text{pyruvate} \xrightarrow{\text{no oxygen}} \text{lactic acid} \xrightarrow{\text{oxygen}} \text{pyruvate}$$

$$\text{NADH} \longrightarrow NAD^+$$

NAD^+ can now participate in the breakdown of another molecule of glucose, allowing glycolysis to continue.

Although animals, like humans, can make up for a temporary lack of oxygen by forming lactic acid, many microorganisms normally extract energy in the absence of oxygen. Such microorganisms are anaerobic. They also undergo fermentation, but the end products depend on the organism. For instance, yeast cells do not produce lactic acid from pyruvate; instead, they form ethanol (alcohol) and carbon dioxide from pyruvate during the oxidation of NADH. This process is called alcoholic fermentation.

$$\text{pyruvate} \xrightarrow{\text{no oxygen}} \text{ethanol} + \text{carbon dioxide}$$

In the following two experiments, you will identify the production of carbon dioxide and ethanol by yeast during alcoholic fermentation. The yeast cultures you will be using have been grown in a solution containing glucose.

Formation of Acetyl-CoA
In the second stage of aerobic respiration (cellular respiration in the presence of oxygen), the pyruvate formed by glycolysis enters the mitochondria. Here, each pyruvate molecule is converted into acetyl -CoA, a molecule with two carbon atoms. The other carbon atom is lost as carbon dioxide. Enough of the coenzyme NAD^+ is reduced to generate five molecules of ATP.

The acetyl-CoA may enter the third stage of cellular respiration, the citric acid cycle, in which more ATP is generated. However, if the cell already has a high level of ATP, it may use the acetyl-CoA to synthesize lipids, thereby storing the energy for later use.

Citric Acid Cycle

The third stage of cellular respiration is the citric acid cycle, also called the Krebs cycle. The citric acid cycle is a series of reactions in which the carbon atoms in acetyl-CoA are converted into carbon dioxide. In addition, three molecules of NADH and one molecule of $FADH_2$ (a coenzyme like NAD^+) are formed. Because oxygen is the ultimate acceptor of electrons carried by coenzymes, the citric acid cycle is also aerobic. In addition to the ATP formed when NADH and $FADH_2$ are oxidized in the electron transport chain, the citric acid cycle generates two molecules of GTP (guanosine triphosphate), which carries the same amount of energy as ATP. In total, 20 molecules of ATP can be produced from one molecule of glucose in one turn of the citric acid cycle.

Electron Transport Chain: Oxidative Phosphorylation

The fourth stage of cellular respiration, the electron transport chain, is the stage in which ATP is produced. The $FADH_2$ and NADH generated in the first three stages transfer their hydrogen atoms (and associated electrons) to oxygen through a series of steps. Oxygen is the most electron-hungry atom in the environment, so this is a "downhill" energy-releasing process, much like the movement of a substance into a cell along a concentration gradient. The mitochondria take advantage of the energy in the attraction between the electrons and oxygen and couple it with the "uphill" process of synthesizing ATP.

4.1 Experiment: Carbon Dioxide Production

In this experiment, you will set up a respirometer, a simple device for measuring the amount of carbon dioxide produced during respiration.

MATERIALS

24-hour yeast culture
Deionized water in 250 mL flask with pipette
5% glucose in 250 mL flask with pipette
5% sucrose in 250 mL flask with pipette
37°C water bath with test tube rack

Per group of two or three students:
Three large test tubes
Three small test tubes
Label tape and markers
Metric ruler

PROCEDURE

1. Practice setting up the respirometer, as shown in *Figure 4.3*.
    a. Fill a small test tube completely with tap water
    b. Invert a large test tube over the small one. Use your finger or the blunt tip of a pencil to push the small tube up until its rim is in contact with the bottom of the large tube.
    c. Quickly invert the tubes so as little water as possible leaks from the small tube. A small bubble will form in the small test tube.
    d. Push filled small test tube into inverted large test tube until its rim touches the bottom of the large tube.
    e. Repeat this procedure until you can make a respirometer with the smallest bubble possible.

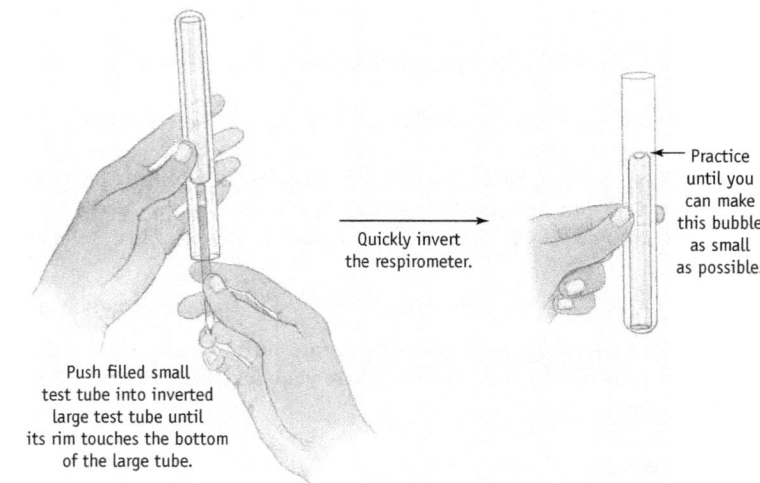

Figure 4.3 Setting Up a Respirometer

2. Use small pieces of tape to label your three small test tubes "control," "glucose," and "sucrose." Recall that sucrose is a disaccharide composed of a glucose molecule linked to a fructose molecule.
3. Mark one small test tube at the 2/3 full level. Use this tube to mark the 2/3 level on the other two small tubes. Fill each tube to this level with the yeast suspensions.
4. Fill each small tube to the top with the appropriate solution. What should the control tube be filled with?
5. Using the technique you practiced, invert each small tube into the empty large tubes so that you have three respirometers.
6. Mark each large tube at the starting level of the air bubble. Place your respirometers onto a 37°C water bath. Allow them to ferment at least an hour while you continue the lab. Check them periodically. As the yeast ferments, the gas bubbles at the top of the small tubes will grow.
    a. After at least an hour, remove the respirometers from the water bath and measure the distance between the final level of the yeast culture and your original mark. Record these measurements on the data sheet.

PRE-FERMENTATION QUESTIONS

Q1. What is responsible for the increasing space at the top of the small tubes?

Q2. What is your hypothesis about the outcome of this experiment?

Q3. Explain your hypothesis.

Post-Fermentation Questions

Q4. How do your results compare with your original hypothesis?

Q5. What difference did you expect between the results in the glucose tube and those in the sucrose tube? What happened? What do you think would happen if you allowed the tubes to ferment longer? Explain your idea.

Q6. Why do you want the original bubble in the respirometer to be as small as possible? What is in this bubble that will interfere with your results?

4.2 Experiment: Ethanol Production

Ethanol is the other product of fermentation. The indicator used in this experiment is Lugol's iodine. When alcohol (clear) and iodine (brownish yellow) are combined in the presence of sodium hydroxide (NaOH), a cloudy yellow precipitate, iodoform, is formed.

MATERIALS

24-hour yeast culture, settled, with pipette
Deionized water in three or four flasks with pipettes
10% NaOH in three or four small flasks with pipettes
70% ethanol in 50 mL flask with pipette
Gloves and goggles for handling NaOH

Per group of two or three students:
Three test tubes
Lugol's iodine in dropper bottle
Test tube rack

PROCEDURE
1. Label your tubes "1," "2," and "3."
   a. Tube 1
      i. Put 1.5 mL of deionized water into the tube.
      ii. Add 1.5 mL of ethanol.
      iii. Slowly add 1 mL of NaOH down the side of the tube and mix by tapping the tube with your finger.
      iv. Slowly add about 10 drops of Lugol's iodine and look for the formation of a cloudy yellow precipitate, iodoform. Record your results on the data sheet.
   b. Tube 2
      i. Carefully pipette 3 mL of the clear solution above the settled yeast cells from the flask into your test tube. Do not mix the yeast culture.
      ii. Slowly add 1 mL of deionized water down the side of the tube and mix by tapping the tube with your finger.
      iii. Slowly add about 10 drops of Lugol's iodine. Record your results on the data
   c. Tube 3
      i. Carefully pipette out 3 mL of the clear solution above the settled yeast cells into the test tube. Do not mix the yeast solution.
      ii. Slowly add 1 mL of NaOH down the side of the tube and mix by tapping the tube with your finger.
      iii. Slowly add about 10 drops of Lugol's iodine. Record your results on the data sheet.

SUMMARY QUESTIONS

Q7. What happened in each tube?

Q8. Which tube was the positive control? Which tube was the negative control?

Q9. What do the results of this experiment tell you about the production of alcohol by yeast?

When you are finished with this lab, dispose of the test tube contents as instructed (remove labels, place empty test tubes into a wash bin).

Cellular Respiration

Data Sheet

Carbon Dioxide Production

Q1. What is responsible for the increasing space at the top of the small tube in the respirometers? _____

_____

_____

Q2. What is your hypothesis about the outcome of this yeast fermentation experiment? _____

_____

_____

Q3. Explain your hypothesis. _____

_____

_____

	CHANGE IN GAS LEVEL (MM)
Control	
Glucose	
Sucrose	

Q4. How do your results compare with your hypothesis? _____

_____

_____

Q5. What difference did you expect between the results in the glucose tube and those in the sucrose tube? What happened? What do you think would happen if you allowed the tubes to ferment longer? Explain your idea. _____
_____
_____

Q6. Why do you want the original bubble in the respirometer to be as small as possible? What is in this bubble that will interfere with your results? _____
_____
_____

Ethanol Production

	CONTENTS	PRESENCE OF ETHANOL?	COMMENTS
Tube 1			
Tube 2			
Tube 3			

Q7. What happened in each of the three tubes? _____
_____
_____

Q8. Which tube was the positive control? Which tube was the negative control? _____
_____
_____

Q9. What do the results of this experiment tell you about the production of alcohol by yeast? _____
_____
_____

Citric Acid Cycle

Q10. Which molecule is the substrate? Which is the product? _____

Q11. Is succinate reduced or oxidized? _____

Q12. Is FADH$_2$ reduced or oxidized in this reaction? What about the oxygen? _____

_____

_____

Q13. What do you expect to happen in Tube 3? How will the results differ from the results of Tube 2? Explain your hypothesis. _____

_____

_____

Q14. Will there be a reaction in Tube 4? How will it differ from the results of Tube 2? Explain. _____

_____

_____

	Contents	**Start time**	**Time until color change (if any)**	**Comments**
Tube 1				
Tube 2				
Tube 3				
Tube 4				

Q15. In the tubes that reacted, you may have noticed that the top surface of the mixture stayed blue. Why is this—what is the top of the mixture exposed to that would keep the methylene blue from reacting with the FADH$_2$? _____

Q16. Was there a reaction in Tube 3? How would it be possible for a reaction to occur in Tube 3? If there was a reaction in Tube 3, how did it differ from the reaction in Tube 2, and why? _____

_____

_____

Q17. Was there a reaction in Tube 4? Did you expect one? How did it differ from the reaction in Tube 2, and why? _____

_____

_____

Questions

Q18. What is the role of the products of alcoholic fermentation in the production of beer? In the production of baked goods? In the production of wine? _____

_____

_____

Q19. Yeasts and many bacteria are **facultative anaerobes**—that is, they make ATP by fermentation in the absence of oxygen, but if oxygen is present, they can undergo aerobic respiration. Why would it be advantageous for them to be able to make ATP by aerobic respiration? Hint: Compare the amount of ATP produced in each process. _____

_____

_____

Q20. Why do you continue to breathe deeply after strenuous exercise? _____

_____

_____

Q21. When your roasting marshmallow catches fire, it burns well, giving off a lot of heat. Where does this energy come from? How is this related to cellular respiration? _____

_____

_____

Lab 5:
Confocal Microscopy
Elizabeth Franklin

OBJECTIVES
1. To gain an understanding of how confocal microscopy works.
2. To view a confocal microscope in action.

Marvin Minsky invented the confocal microscope in 1955. He wanted to study neurons and their networks and the technology at that time lacked the resolving power. Traditional forms of microscopy could not distinguish the finer details of complex samples, such as a cell membrane or neuronal tissue. The key to confocal microscopy's resolving power is the **elimination of any out of focus light**. Without the blurring from the out of focus light, much finer details could be observed. Only one single point is illuminated giving a resolving power of up to (currently) ~80 nanometers

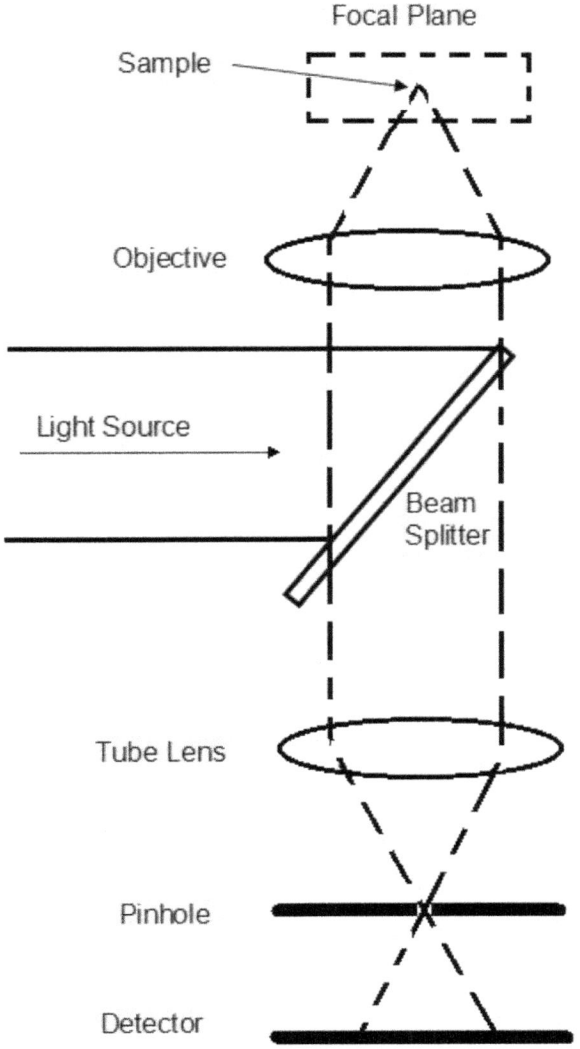

The key principle in the design of the confocal microscope is the **pinhole**. The pinhole is what allows for the elimination of the glare from any out of focus light. *Figure 5.1* illustrates the basic design of the confocal microscope and the parts involved: sample, objective, light source, tube lens, pinhole, and detector. The light source, commonly a laser, comes in and is reflected to illuminate the sample. The sample is in the **focal** plane which is **con**jugate to the plane of the pinhole in front of the detector. When the light hits the pinhole, only the light that illuminates the particular point of choice is allowed through to the detector.

Where do you think the term confocal comes from?

Confocal microscopy has provided biologists with a way to observe structures in 3-D and real-time. Confocal microscopy uses a computer to control the microscope and run the software to process the images being taken. Many single images are taken over a period of time and with enough images, a 3-D image can be built using the right software, like slices layered on top of each other until a 3-D image forms. Confocal microscope images can all provide real time information about the sample. An example would be observing the cellular process of endocytosis. You could observe the steps, but it would be like watching an old movie. Each frame tells what is occurring every few seconds, but the images are not smooth or continuous.

Figure 5.1 How a confocal microscope works

There are currently several forms of confocal microscopy: confocal laser scanning microscopes, spinning-disk confocal microscopes, dual spinning-disk confocal microscopes, and programmable array microscopes. Each of these varies in design, but all use the concept of pinholes to eliminate unfocused light.

In the past we have observed adenocarcinoma (breast cancer) cells. Check with your instructor to see what you will be observing today.

POST-LAB QUESTIONS

Q1. In your own words describe how confocal microscopy works. Be sure to use key terms.

Q2. What did you observe today in lab? Describe how the confocal worked, what software was used, and how did the software work?

Q3. If you could look at anything you wanted under the confocal microscope, what would it be and why?

Q4. Many advances have improved upon Minsky's original device. Research the history of the confocal microscope, pick a particular advancement and briefly elaborate on the advancement and why it is important.

Lab 6:
Photosynthesis: Capturing Energy
Carolyn Eberhard

OBJECTIVES
1. Identify the pigments involved in photosynthesis by paper chromatography.
2. Give the function of chlorophyll and the accessory pigments
3. Determine the wavelengths most useful for photosynthesis by determining the absorption spectrum of a pigment extract.
4. Distinguish between the electron transport and carbon fixation reactions of photosynthesis.
5. Demonstrate the effect of changing the level of light on electron transport as measured by oxygen production.
6. Demonstrate the effect of changing the level of light on electron transport by measuring reduction of a dye in the Hill reaction.
7. Explain why stimulating carbon fixation affects electron transport as measured by oxygen production.
8. Describe an experiment designed to measure the extent of carbon fixation as a function of light.

Key Words
Chloroplast Electron transport
Electromagnetic radiation
Chemiosmosis
Photosynthetic pigment
Photophosphorylation
Photosystems I and II
Carbon fixation
(consult text index for page references)

INTRODUCTION

The overall process of photosynthesis can be understood as two closely linked sets of reactions. The **light-dependent reactions** occur in the thylakoid membranes of the chloroplast, where light-driven electron transport produces NADPH and sets up a proton gradient for ATP synthesis by **chemiosmosis**. ATP and NADPH are then used in the **light-independent reactions**. These occur in the fluid stroma regions of the chloroplast, where **carbon fixation** is carried on by the enzymes of the **Calvin cycle** to convert CO_2 to carbohydrates such as glucose. The whole process is diagramed in *Figure 6.1*. (See Gallery Photo 7.)
The driving force of photosynthesis is light absorption by the photosynthetic pigments of **Photosystems I and II**. Only the wavelengths of light that are actually absorbed can be used.

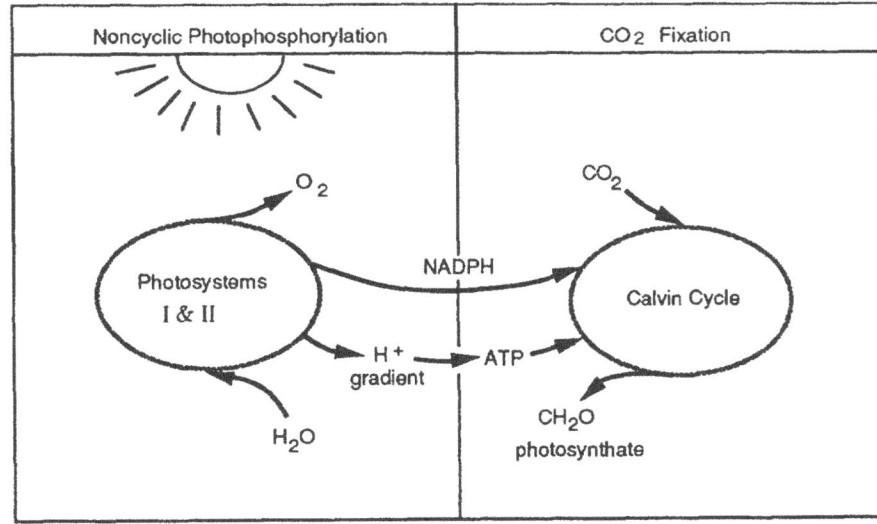

Figure 6.1 ATP synthesis by chemiosmosis depends on a hydrogen ion gradient set up as a result of electron transport. The Calvin cycle is tightly coupled to electron flow and quickly stops if light is not available.

(Modified from Cornell University Laboratories.)

Light energy drives the noncyclic flow of electrons from water molecules, where they are obtained when water is split to oxygen and hydrogen, to NADP+ along a chain of carriers. The flow also sets up an H+ gradient that is used for the synthesis of ATP by chemiosmosis. The entire process by which light energy is converted to chemical energy in the form of NADPH and ATP is **noncyclic photophosphorylation**. The flow of electrons is noncyclic from the source in water molecules to the hydrogen carrier NADPH. The energy to drive the flow "uphill" comes from light (photo-). ADP undergoes phosphorylation carried on by ATP synthetases in the thylakoid membrane for which energy is supplied by a proton (H+) gradient. Sometimes ATP alone is made by cyclic photophosphorylation, but this process is much harder to study because O_2 and NADPH are not produced.

In the light-independent reactions of the Calvin cycle, the chemical energy of NADPH and ATP is used to reduce CO_2 to carbohydrate, an "uphill" endergonic process. Carbon fixation begins with CO_2 uptake into an organic molecule, a reaction catalyzed by the enzyme ribulose bisphosphate carboxylase. Eventually, low-energy CO_2 is converted to high-energy carbohydrates. Actually, these reactions also depend on light when they occur in the intact plant, because they require NADPH and ATP. As long as noncyclic photophosphorylation continues, sufficient NADPH and ATP will be available for the Calvin cycle reactions, CO_2 will be fixed and carbohydrate will be synthesized. If light is cut off or electron flow is blocked, ATP and NADPH will quickly be used up and the Calvin cycle will stop operating. Thus, in intact photosynthetic systems, light is necessary for the Calvin cycle because it is so tightly coupled to the synthesis of ATP and NADPH.

In the light-independent reactions of the Calvin cycle, the chemical energy of NADPH and ATP is used to reduce CO_2 to carbohydrate, an "uphill" endergonic process. **Carbon fixation** begins with CO_2 uptake into an organic molecule, a reaction catalyzed by the enzyme ribulose bisphosphate carboxylase. Eventually, low-energy CO_2 is converted to high-energy carbohydrates. Actually, these reactions also depend on light when they occur in the intact.

PRELAB QUESTIONS

Q1. Write out the balanced equation for the synthesis of 1 glucose molecule from 6 carbon dioxide molecules and 12 water molecules.

Q2. Write the starting materials and circle the oxygen molecules that will produce the waste product O_2.

6.1 Experiment: Analyzing the Chromatogram

The number of spots that you observe will depend on the exact condition of the extract that you used. These are the major pigments in the extract:

Chlorophyll *a*: a blue-green pigment
Chlorophyll *b*: a yellow-green pigment
Carotenes: yellow-orange accessory pigment
Xanthophylls: yellow accessory pigments

MATERIALS
Chromatogram from Experiment 6.1
Chromatography Jar and Cover
Solvent

PROCEDURE
1. Outline each visible spot with a pencil, and note its color on the chromatogram.
2. Make a dot with your pencil in the center of each spot. If the spot is irregular or band-shaped, use your judgment as to where the approximate center might be if it were a round spot.
3. Carefully measure the distance from the origin to the solvent front, and record it in a table. Measure the distance from the origin to the dot in the center of each spot observed.
4. Record it on the chromatogram and in *Table 6.1*.
5. Calculate the R*f* for each spot, and describe its appearance under visible light.
6. If you can, identify the major pigments, and record your R*f* values for them in the *Table 6.1*.

Pigment molecules in the faster spots will be more soluble in the hydrophobic solvent; they will be smaller and have less affinity for the hydrophilic paper. For instance, chlorophyll *a* and chlorophyll *b* are identical except for one functional group:

$$\text{Chlorophyll } a: \text{R}-\text{CH}_3$$

$$\text{Chlorophyll } b: \text{R}-\overset{\overset{\displaystyle O}{\|}}{\text{C}}-\text{H}$$

Table 6.1 Data from Chromatography

Spot	cm moved	R_f (cm/cm solvent)	Color (Visible)	Color (UV)	Identification
Solvent front		—	—	—	—
Fastest spot					
Slowest spot					

POSTLAB QUESTIONS

Q1. Which molecule would you expect to travel faster in your chromatography?

Q2. Why does polarity affect the rate of diffusion of molecules down a gradient?

6.2 Experiment: The Absorption Spectrum

Chlorophyll *a* and *b* pigments are located in the thylakoid membranes of the chloroplasts where they normally would absorb light, become chemically excited, and pass electrons to the electron carriers of photosystems I and II. When they are free, however, as they are on the chromatography paper, there are no molecules to accept the electrons, and the UV light-excited molecules release their energy as photons of red light during fluorescence. Only the molecules of certain compounds are able to give up the energy as light. Those of other compounds lose it in the form of heat.

Only the light that is absorbed by chloroplasts can be used in photosynthesis. As you have already seen, the pigments of a green plant look green to the eye because they permit green light to pass through, but absorb the red and blue light. The particular wavelengths of light that are absorbed by a certain substance form a pattern called its absorption spectrum. Illuminating a solution of the substance with each wavelength of
light in turn and measuring the absorption in each case determine the spectrum. Using visible light, you will determine the visible spectrum for a mixture of pigments extracted from chloroplasts. The spectrum can then be plotted as a graph of absorbance versus the wavelength or color of light used.

In this experiment, you will have to be careful not to spill the solvent on or into the spectrophotometer because it will dissolve the plastic.

MATERIALS
Spectrophotometer
80% Acetone Chlorophyll
Extract Graph paper

PROCEDURE
1. Turn on the spectrophotometer to warm up for a few minutes. In the mean time, be sure that you are familiar with the basic procedure for this instrument.
2. Fill a spectrophotometer cuvette halfway with 80% acetone to use as your blank.
3. Fill a second tube halfway with diluted **Chlorophyll extract**.
4. Set the wavelength at 400 nm. Insert your blank, and blank the instrument
5. Insert the tube containing chlorophyll, and take your first reading. It should be in the vicinity of 0.8. If it is not, ask your instructor for help. Record the absorbance in a table.
6. Change the wavelength to 405, reset the spectrophotometer, and take your second reading; enter the absorbance in your table. Continue your readings every 5 nm until the readings have become quite low. Then switch to every 10 or 20 nm.
7. When the absorbance starts to rise, switch back to take readings every 5 nm, and continue until you reach 700 nm (you can use *Table 6.2* to collect your data).
8. Plot your data on graph paper, and label each absorption maximum with its corresponding wavelength.

*The purpose of taking the spectrum is to get the shape as accurately as possible, especially around the regions of greatest absorbance (low 400s and high 600s). When the absorbance rises and then falls again, the peak is called the absorption maximum, and the wavelength at which it occurs is used as a characteristic for identifying a compound. If you are not sure where at least two maxima are in your spectrum, go back to the regions of high absorbance and take some more readings.

The zero and absorbance will change each time you adjust the wavelength, so you will have to re-blank the spectrophotometer with each change of wavelength.

PRELAB QUESTIONS
Q1. Why is 80% acetone used as a blank sample rather than pure water?

Table 6.2 Absorption Spectrum of Photosynthetic Pigments

Wavelength	Absorbance	Wavelength	Absorbance	Wavelength	Absorbance
400 nm					

6.3 Experiment: The Hill Reaction

In 1937, Robin Hill showed that chloroplasts continue to carry out electron flow and oxygen production in the absence of CO_2 as long as they are provided with an electron acceptor (see *Figure 6.2*). Hill used a substance that becomes colored as it is reduced, so that the rate of photosynthesis could be measured by observing the rate of absorption, you will use a blue dye nicknamed DCIP or DPIP (2,6- dichlorophenolindophenol), which is blue when oxidized and becomes colorless after it has been reduced:

Figure 6.2 The Hill reaction. In this experiment, an electron acceptor, DCIP, is provided to pick up electrons from Photosystem 1. This allows noncyclic electron transport and water splitting to continue in the absence of the Calvin cycle. DCIP is observed to turn blue as it is reduced.

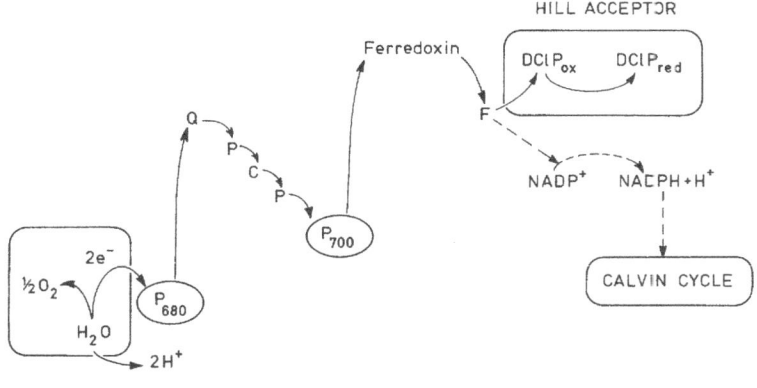

$$DCIP_{ox} + 2e^- + 2H^+ \longrightarrow DCIP \cdot H_{2red}$$

The experimental tube will contain "living" chloroplasts, buffer, the blue dye, and will be held in the light.

PROCEDURE
1. Set the wavelength of the spectrophotometer at 620 nm.
2. Obtain three spectrophotometer tubes and label them #1, #2, and #3.
3. Add these reagents in the order given (left to right) in *Table 6.3*.

Table 6.3 Components for the Hill Reaction

	Chloroplasts	**Cold Buffer**	**Distilled Water**	**DCIP**
Tube #1: blank	0.5 mL	3.0 mL	1.5 mL	—
Tube #2: reaction	0.5 mL	3.0 mL	0.5 mL	1.0 mL
Tube #3: control*	—	3.0 mL	1.0 mL	1.0 mL

* A tube identical to tube #2 but held in the dark would be an equally good control but would use up the chloroplasts.

4. Cover the tubes with Parafilm and mix the ingredients by inverting the tube several times.
5. Immediately cover the tubes with aluminum foil, and keep them covered except when they are being measured.
6. Adjust the spectrophotometer to absorbance, insert tube #1, and adjust the absorbance to 0. Insert tube #2, and record its absorbance in a table. Cover the tube and wait 3 min.
7. Again read the absorbance of tube #2.
8. Hold your tube at the appropriate position for exactly 30 sec. Cover the tube.
9. Read the absorbance; cover the tube again.
10. Hold the tube in the light for another 30-sec period, and again read the absorbance. Continue until you have 12 readings. Since the experiment is meaningless without the control, read the absorbance of tube #3 (no chloroplasts).
11. Hold the tube for 3 min at the same light intensity that you used in the experiment.
12. Read the absorbance again.
13. Graph your results on graph paper as percent DCIP reduced as a function of seconds exposed to the light.

% DCIP reduced = (change x 100) / Initial

Table 6.4 Tube #2 Reaction Data

Seconds of Light	Absorbance	Change in Absorbance	% DCIP Reduced
0 (initial)			
0 (after 3 min dark)			
30			
60			
90			
120			
150			
180			
210			
240			
270			
300			
330			
360			

Table 6.5 Tube #3 Control Data

Seconds of Light	Absorbance	Change in Absorbance	% DCIP Reduced
0			
180			

Lab 7:
Quantification of Cellular Protein, Part I
Jason Wolfe, Dr. Maria R. Davis & Dr. Debra Moriarity

OBJECTIVES
1. Demonstrate the principles of colorimetry, serial dilution, biochemical assays, and cell protein content.

INTRODUCTION

It is often necessary to determine the concentration of a protein in dilute solutions. A sensitive yet simple colorimetric assay was developed by Bradford (Analytical Biochemistry [1976] 72 p.248-254). This assay uses the fact that the dye Coomassie Brilliant Blue, binds directly with a protein solution. Upon binding, the dye changes color, absorbing light at a different wavelength. The absorption maximum shifts from 465 to 595 nm. With a spectrophotometer, it is possible to quantify color by measuring the amount of light absorbed by the solution. Over a narrow range of 10 – 100 μg protein, absorption at 595 nm is linear.

A standard curve provides a reference for measuring the amount of protein in a solution of unknown concentration. It is established by measuring the absorption of known concentrations of protein (typically in duplicate) between 10-100 μg. The standard curve for today's exercise will be constructed using a solution of bovine serum albumin (BSA) of known concentration.

The absorbance of the unknown sample can then be compared with that of the standard curve. The problem with this technique is that there is no way to determine, ahead of measurement, whether the unknown sample actually falls within the 10 and 100 μg range in which the determination of protein concentration is accurate. The unknown sample could have more or less protein than this. It is therefore necessary to perform serial dilutions of the unknown sample. Since the linear portion of the standard curve, in which the measurement is accurate, extends over an order of magnitude (10 – 100 μg), the best approach involves preparing serial dilutions that differ by one order of magnitude in protein concentration.

The measurements for construction of the standard curve and determination of the concentration of the unknown dilution must be done at the same time. This is to reduce experimental error due to different developing times for the Bradford reagent.

7.1 Experiment: The Standard Curve

MATERIALS
Bovine Serum Albumin 1mg/mL (0.5mL)
0.01% SDS / 0.1mM EDTA (15mL) Bradford Reagent (50mL)
100mL beaker for Bradford Reagent (1) Unknown protein (3 x 1mL)

Test tubes (10) Plastic
Cuvettes (10)
Pipets (3 x 2mL, 1 x 10mL) Spectrophotometer set at 595nm Graph Paper

PROCEDURE
1. Obtain a solution of protein whose concentration is known. Bovine serum albumin (BSA) has been prepared in SDS/EDTA at a concentration of 1 mg/mL. Dilute 0.5 mL of BSA with 4.5 mL of 0.01% SDS / 0.1 mM EDTA to yield 5 mL at 100 μg/mL.
2. Label a set of test tubes from 1-6 using a black Sharpie marker or pen on tape. Following the chart below, fill the tubes sequentially with BSA, SDS/EDTA and Bradford reagent (see note). Bradford Reagent contains a dye, Coomassie blue, which binds to protein. The dye/protein complex produces a blue color whose absorbance is directly proportional to the protein concentration.
3. Plot your data on graph paper, μg/mL protein on the X axis and A_{595} on the Y axis. If a linear graph is not obtained, repeat the assay being more careful with pipetting.

NOTE: Add Bradford reagent to all 10 tubes at once, only after having done the Standard and Serial Dilutions. Show tubes to instructor to obtain reagent.

It is important to mix the tubes rapidly and thoroughly immediately after the dye is added, one tube at a time. Because the Bradford reagent contains phosphoric acid, avoid contact with the mouth or skin.

Table 7.1   Data table for Standard Curve

Tube No.	100 µg/mL BSA (mL)	SDS/EDTA (mL)	Bradford Reagent (mL)	Protein (µg/ml)	A_{595nm}
1	0.0	1.0	2		
2	0.2	0.8	2		
3	0.4	0.6	2		
4	0.6	0.4	2		
5	0.8	0.2	2		
6	1.0	0.0	2		

7.2 Experiment: Assay of the Unknown

You will determine the concentration of an unknown solution of BSA in SDS/EDTA. Dealing with unknowns, determining as much information about them as possible is what makes science so much fun. However, the enjoyment factor is nearly canceled out by the frustration level if the investigation is not carried out in an orderly fashion. For example, the way to ascertain the concentration of BSA in the unknown is to compare its absorbance to the standard curve created in the first part of this exercise. You might guess that a five-fold dilution might do the trick. However, if the A_{595} does not fit the curve, it will be necessary to dilute the sample again. If the absorbance reading still does not fit, you would have to try yet another dilution. This "shot in the dark" technique not only takes a lot of time but uses up quite a bit of sample and provides plenty of opportunity for error.

The most efficient way to determine the concentration of an unknown is by using the strategy of the serial dilution. By serial diluting the unknown by an order of magnitude (10-fold) each time, you will be guaranteed to have one tube with an absorbance value that fits the standard curve. Once the concentration of unknown is determined for a particular tube, it is only a matter of a simple "back calculation" to determine the amount of protein in the original sample.

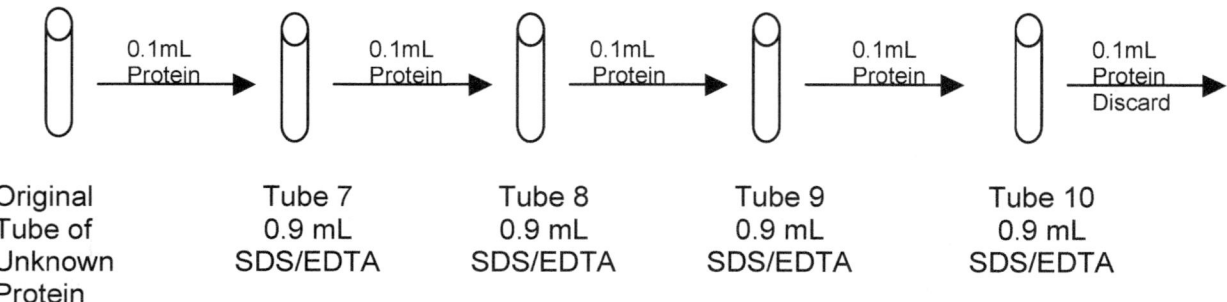

Figure 7.1 A dilution series.

PROCEDURE
1. Label a set of test tubes 7-10 and pipet 0.9 mL of SDS/EDTA into each tube. To the first tube add 0.1 mL of unknown for a total of 1.0 mL. Tube 7 now contains a 10-fold dilution of the unknown or an order of magnitude less protein than the original tube. Mix the tube gently by pipetting up and down or vortexing gently; try to avoid creating bubbles. This action also rinses the walls of the tube to remove any excess protein. Why is this important?
2. Transfer 0.1 mL from tube 7 to tube 8; mix gently. The second tube now contains a 10-fold dilution of the protein in the first tube, or 2 orders of magnitude (10 x 10) less protein than the original ($C_o \times 10^{-2}$).
3. Now transfer 0.1 mL from tube 8 to tube 9, mix greatly. Transfer 0.1 mL from tube 9 to tube 10.
4. How many orders of magnitude less than the original do you have in this last tube? Express it as a fraction. Express it also as an exponent. What volume of solution would you have ended up with if you had made this dilution directly from the original sample? In other words, starting with 0.1 mL of unknown, how much SDS/EDTA would be necessary to achieve this same degree of dilution without using the serial dilution technique?
5. Remove 0.1 mL from tube 10 and discard it (into the sink). What is the reason for this?
6. Now add Bradford reagent to all 10 tubes. Full color development should occur in 5 minutes. Assay the tubes with the spectrophotometer within 1 hour. Use the tube #1, the BLANK, to set the absorbance at 595 nm to zero. Measure the A_{595} of each of the other tubes. Record the values in the chart above.
7. Chose the absorbance value that fits the Standard Curve appropriately and graphically determine the protein concentration for that tube. Show your graph to the instructor.
8. Calculate the concentration of the "Unknown Protein" in the "Original tube": Use the tube with the "best fit" to the standard curve for the back calculation. For example, if the amount of protein in tube 8 is 35 µg (in a 1 mL volume), then the concentration in the original tube is:

$$10 \times 10 \times 35 = 3500 \text{ µg/mL} = 3.5 \text{ mg/mL}$$

Table 7.2 Data Table for Serial Dilution

Tube No.	Unknown Protein (mL)	SDS/EDTA (mL)	Bradford Reagent (mL)	A_{595nm}	Order of Magnitude (Dilution)	Dilution Factor	Mark (_) for the tube that fit the O.D range of the Std Curve	Conc. of protein in the Marked (_) tube (µg/mL)
7	0.1	0.9	2					
8	0.1	0.9	2					
9	0.1	0.9	2					
10	0.1	0.9	2					

Note: If you accidentally record transmittance data instead of absorbance, use the following equation (The Lambert-Beer law).

$$A = \log(1/T) = -\log(T)$$

The Use of a Spectrophotometer

Philip Stukus

A spectrophotometer is an instrument that measures the amount of light that is absorbed or transmitted when passed through a sample. When the ab- sorbance of light is measured, the instrument is called an absorption spectrophotometer. Many biologically important molecules absorb specific types of radiant energy (light of differing wavelengths). In microbiological appli- cations, a beam of light passes through a liquid suspension containing bacteria. The larger the population of bacteria the greater the amount of light scattering. As a result, less light passes through the sample.

The basic design of a spectrophotometer is shown in *Figure 7.2*.

A source of white light is used as the light source. The light is focused on a dispersion device, usually a prism, to separate the light into individual bands of radiant energy. Each wavelength of light is focused through a narrow slit and then passes through the sample to be measured. The sample is usually held in a glass or plastic tube called a cuvette. After the light passes through the sample it strikes a light-detecting photoelectric cell. When the transmitted light strikes the photoelectric cell, an electric current is generated which is proportional to the transmitted light energy. The current is dis- played on a meter or by digital readout. Most instruments display the amount of light transmitted (% transmission) and the amount of light absorbed.

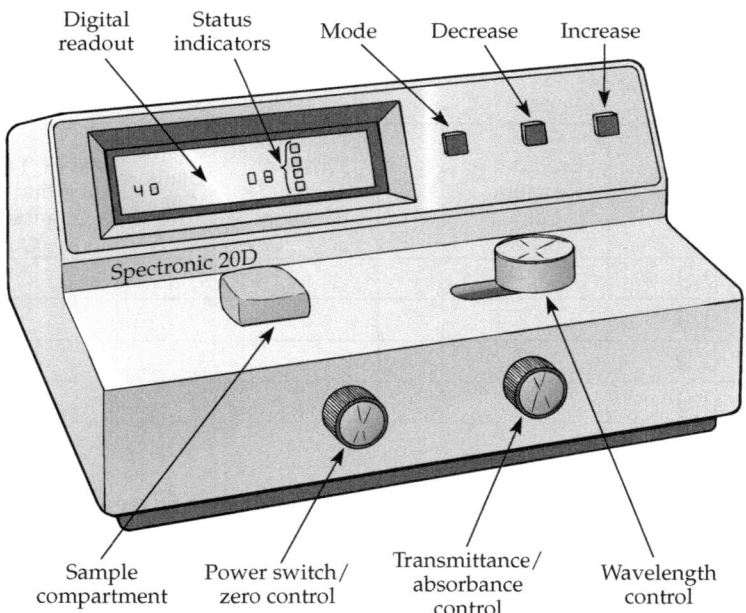

Figure 7.2 A Spectrophotometer

Procedure for Measuring Bacterial Turbidities

This procedure describes measurements taken with a Milton Roy Spectronic 20 or 21. The wavelength control knob adjusts the angle of deflection of the light source in order to obtain a specific wavelength of light. The power switch also serves as the zero control knob and is used to turn the instrument on and off and to set the %T reading.

1. Turn on the spectrophotometer by turning the power switch knob clock-wise. Allow the instrument to warm-up for 15 min prior to use.
2. Set the desired wavelength with the wavelength control knob.
3. On the Spectronic 20 adjust the meter to 0% T with the zero control knob.
   *Notice that there is a mirror directly behind the needle. When taking a reading, be sure that the needle is directly in line with its reflection in the mirror.
4. With the Spectronic 21, select the operating mode for either absorbance or % transmission (%T) and set the sensitivity switch to low.
5. Place a tube or cuvette containing the uninoculated culture medium into the sample compartment.
   *This is a control that takes into account the light scattering or absorbance of the cuvette and the culture medium. We are interested in only determining the amount of light scattered by the bacteria present in the culture medium. Be sure that the cuvette or tube has been wiped clean to remove any liquid, dust, or fingerprints.
6. Adjust the meter to 100% T (0 Absorbance) with the transmittance/absorbance control.
   *Be sure that the inoculated culture tube or cuvette has been wiped clean and that it has been placed securely in the sample compartment.
7. Close the lid and read the absorbance.
8. When all readings have been completed, turn off the instrument by turning the power switch counterclockwise until it clicks.

Procedural Hints

1. Inaccurate readings very often can be attributed to
   (a) Improper zeroing of a blank solution. (Remember, the blank is the uninoculated tube of the particular culture medium used. If multiple media are used, each has to be zeroed separately.)
   (b) Cuvette or tube not being placed securely in the sample compartment.
   (c) Wavelength not being set properly. (When there are multiple users of a spectrophotometer, each user may be taking readings at different wavelengths.)
2. If any culture is spilled in the spectrophotometer, notify your lab instructor immediately.

Lab 8:
Quantification of Cellular Protein, Part II
Dr. Debra M. Moriarity

OBJECTIVES
1. Learn the use of the hemocytometer in the counting of cells.
2. Assay mammalian cells in tissue culture for protein content.
3. Correlate the protein content with cell number.

*The cells we will be examining today are a cell line (MCF-7) that was established from a tumor of a patient with breast cancer. The cells are maintained in cell culture and were purchased from the American Type Culture Collection.

INTRODUCTION

This lab consists of two sections to be performed in one lab session.
> Part 1: The cells will first be removed from the tissue culture plates with trypsin, washed, diluted and then counted in a hemocytometer.
> Part 2: An established number of cells will be solubilized in a buffered detergent solution. Serial dilutions will then be made and the protein content assayed.

These cells have been growing in culture for several days and should be close to **confluency**. That is, they should be covering the plate and touching one another. To remove the cells from the tissue culture flask, it will be necessary to treat them with the proteolytic enzyme **trypsin** in a solution with ethylenediamine tetra- acetic acid (EDTA) that chelates (traps) calcium ions. This solution will substantially dissolve their attachments both from each other (the Ca^{2+} necessary to maintain association between cell-surface proteins will be removed by the EDTA) and from the extracellular matrix protein they have secreted onto the flask. Such treatment will allow the cells to be removed from the flask by gentle **trituration** using buffer and a pipet. Sterility is no longer a concern with these cells, but cleanliness (to avoid introducing foreign proteins) is important. After removal into a plastic tube, the cell suspension should be mixed gently and three or four 10 µl samples taken for counting in a hemocytometer. If there are three members of the group make one count each. If there are two members then each should make two counts.

The hemocytometer has two separate counting areas. Each area has nine large squares. In the corners are large squares divided into 16 small squares. On the sides are large squares with 20 smaller rectangles; the large square in the center has 25 small squares, each subdivided into 16 smaller squares. This sounds complicated, but using the device is not. A special coverslip is supported above the grid such that the volume above each large square is equal to precisely 0.1 µl. If 30 cells were located in one large square then your sample would contain 30×10^4 cells/mL. Typically, the area contained by the four large outer squares and the one large (subdivided) central square are examined and the cells therein are counted. Multiplying the number of cells obtained from all five square by 2000 gives the number of cells suspended in each mL of media.

8.1 Experiment: Preparation and Counting of Tissue Culture Cells

Side A = Wall of flask on which cells are spread
Side B = Wall of flask opposite to cells

MATERIALS
Compound microscope
Hemocytometer
Hank's Buffered Salt Solution (HBSS) –
    (3mL – warm, 5mL – ice cold)

0.9% Saline Solution supplemented with 0.02%
Triton X-100 detergent (10mL) Trypsin-EDTA
solution (3mL)
Flasks of breast adenocarcinoma cells
Pipets (2 x 10mL) Micropipettes

Test tubes (11)
Cuvettes (11)
1.5mL Eppendorf tubes (1)
Conical Tubes (1 x 15mL)
Bradford Reagent (22mL)

Bovine Serum Albumin (BSA) standard (1.5mL - 200µg/mL)
Spectrophotometer
2mL of 0.4% Trypan Blue

PROCEDURE
1. Pour the media off the cells (from Side B).
2. Using a clean pipet, add a few (2-3) mL of warmed (37°C) Hanks Balanced Salt Solution (HBSS) buffered to pH 7.35 with 25 mM HEPES, (on Side B). Gently rock the solution over the monolayer of cells (Side A) to rinse the serum-containing medium from the cells (serum has components that inhibit trypsin).
3. Remove the solution from the flask completely (pour out from Side B).
4. Using a clean pipet add 2.0 mL of trypsin-EDTA in HBSS (at Side A) and let sit at room temperature for 1-2 minutes.
5. Add 5.0 mL of ice cold HBSS (at Side A) using a clean pipet and suspend the cells by CAREFULLY pipeting the solution up and down over the cell monolayer (Do this gently or you will lyse the cells). Alternately, you can rock the flask gently until the layer of cells comes loose.
6. Transfer the cell suspension to a clean tube and keep on ice.
7. Determine the number of cells per unit volume using a hemocytometer.
8. Use of a hemocytometer
    a. Place the hemocytometer with coverslip in position on the microscope stage. b. Add 0.1 mL 0.4% trypan blue to 0.9 mL cells in a 1.5 mL microcentrifuge tube and mix gently by inverting twice.
    b. Place a drop of evenly suspended cells at the edge of the coverslip. The drop will be drawn into the measuring field by capillary action. If the drop of fluid does not fill the entire field, the whole sample must be discarded (start over).
        i. If the volume of the drop is too great and it flows into the gutters, the whole sample must be discarded (start over).
        ii. If the drop is held too long before applying it to the hemocytometer field, the whole sample must be discarded (start over).
    c. Count the designated outer squares and the inner square with the viable cells only. (Note: The dark blue stained cells are non-viable or dead.)
        i. If cells impinge on the lines that designate the upper or left hand edge of a (big) square then they are NOT to be counted.
        ii. If the cells impinge on the lines that designate the lower or right hand edge of a (big) square, then they ARE to be counted.
    d. Averaging the Cells
        i. The result of the count of the five (big) squares should be multiplied by 2000 to obtain the cell number per milliliter.
        ii. The results of the three or four independent counts should be averaged to find the "true" number of cells per milliliter.
        iii. Your number will probably be something on the order of a million cells per milliliter.
        iv. Note: the Trypan Blue will add a 10% dilution effect to your sample that can be easily corrected for in your calculations.

8.2 Experiment: Protein Assays

PROCEDURE

1. Take a fixed number of suspended tissue culture cells (say 3 million ~ 3mL) and pellet the cells in a centrifuge for 5 minutes. These are cells that you are going to lyse with detergent for the protein assay.
2. Decant the supernatant and resuspend the pellet in 0.9% saline solution with 0.02% Triton X-100 at a concentration of 1 million cells/mL. Agitate them over the course of a few minutes. This will dissolve the membrane of the cell and most organelles and release the protein contents.
3. Make a standard curve of BSA in 0.9% saline solution with 0.02% Triton X-100. **Do not add** the Bradford Reagent until you have prepared your serial dilution samples as described in the later steps.

Table 8.1 Data Table for BSA and Absorbance

Tube	200 µg/mL BSA Solutio	0.9% saline w/	Bradford Reagent	*BSA Concentration (µg/mL)	A_{595nm}
A	0 mL	0.5 mL	2 mL		
B	0.05 mL	0.45 mL	2 mL		
C	0.125 mL	0.375 mL	2 mL		
D	0.25 mL	0.25 mL	2 mL		
E	0.375 mL	0.125 mL	2 mL		
F	0.5 mL	0	2 mL		

4. Prepare a serial dilution of your solubilized cell suspension. Label the tube in which you dissolved your cells as tube #1. You will assay this, undiluted. Use fresh tubes.
   a. Label a set of four tubes (2-5) and pipet 1.35 mL of 0.9% saline solution with 0.02% Triton X-100 into each.
   b. Into tube #2, add 0.150 mL of cell suspension from tube #1. Mix well.
   c. Into tube #3, add 0.150 mL of cell suspension from tube #2. Mix well.
   d. Into tube #4, add 0.150 mL of cell suspension from tube #3. Mix well.
   e. Into tube #5, add 0.150 mL of cell suspension from tube #4. Mix well.
5. You will now take one 0.5 mL sample from each of tubes #1-#5 and place them into fresh, labeled tubes (#1a, #2a etc).
6. Add 2 mL of Bradford reagent to the tubes of the standard curve (A-F) as well as the tubes of your serial dilution.
7. Full color development should occur in five minutes. Assay the tubes with the spectrophotometer (absorbance 595 nm) within an hour. Use tube A as a blank.
8. Plot the data from the standard curve on graph paper with micrograms of protein on the X axis and A_{595} on the Y axis. Use averages from the duplicates for the Y axis data. Show this to your instructor. If a linear graph is not obtained then repeat the assay, being more careful with the pipetting.
9. If a linear graph is obtained, see which sample dilution falls along the graph. Determine the protein content per cell with this information.

Lab 9:
Kingdom Fungi
Alison Morrison-Shetlar

INTRODUCTION

Upon first observation of a mushroom or toadstool in your yard, you might think that you are looking at a plant. Indeed, these and most fungi are multicellular organisms made up of cells that have cell walls—a typical plant characteristic.

However, on closer inspection, you would notice that the fungi are lacking chlorophyll, leaving them unable to produce their own food (although many plants lack chlorophyll, as well). Fungi are either decomposers or parasites, and they have extracellular digestion and absorption of organic nutrients. Another unique characteristic of the fungi is that the body of a fungus, or mycelium, is made up of many branching and unbranching filaments, called hyphae. Also, most fungi produce both sexual and asexual spores, depending on the environmental conditions. Other methods of asexual reproduction are fragmentation or budding.
The following is an outline of the phyla of fungi:

Phylum Zygomycota
1. Characteristics: zygomycetes; lack cross walls; produce non-motile asexual spores that develop inside a sporangium (pl. sporangia) found on hyphae called sporangiophores; form sexual spores (zygospores) from the conjugation of 2 special hyphae.
2. Examples: bread molds and related forms, Rhizopus, Phycomyces.

Phylum Ascomycota
1. Characteristics: ascomycetes ("sac fungi"); usually reproduce asexually by forming spores (conidia) at the tip or sides of special hyphae (coni- diophores); reproduce sexually by forming ascospores (usually 8, sometimes 4) within an ascus (pl. asci); asci may be formed into a fruiting structure called an ascocarp), form mutualistic associations with the roots of most trees (mycorrhizae), fungal components of most lichens.
2. Examples: most yeasts (Saccharomyces) and molds, morels, truffles, powdery mildew, Dutch elm disease, ringworm (humans).

Phylum Basidiomycota
1. Characteristics: basidiomycetes ("club fungi"); reproduce asexually by budding, fragmentation of mycelium, or forming conidia; reproduce sexually by forming a club-like basidium (pl. basidia) which bears haploid spores (usually 4) on its surface; have dikaryotic stage.
2. Examples: mushrooms, plant rusts and smuts, puffballs, toadstools, shelf and bracket fungi.

Phylum Deuteromycota
1. Characteristics: deuteromycetes ("imperfect fungi"); asexual reproduction by formation of conidia; do not have sexual reproduction (at least none that we know of).
2. Examples: Aspergillus, Penicillium (some produce the antibiotic penicillin, others give Camembert and blue cheeses their flavor), athlete's foot fungus, Candida albicans (causes thrush in the mouths of infants); Microsporum.

The following page contains detailed diagrams for the life cycle of zygomycetes and mushrooms (*Figures 9.1 and 9.2*).

Most zygomycetes, including the common bread mold, are saprophytic and their vegetative hyphae do not possess septa. They produce sporangia, which are capsules where haploid spores develop from the process of mitosis. A spore is usually haploid (has half the number of chromosomes) and divides mitotically to produce a multicellular organism.

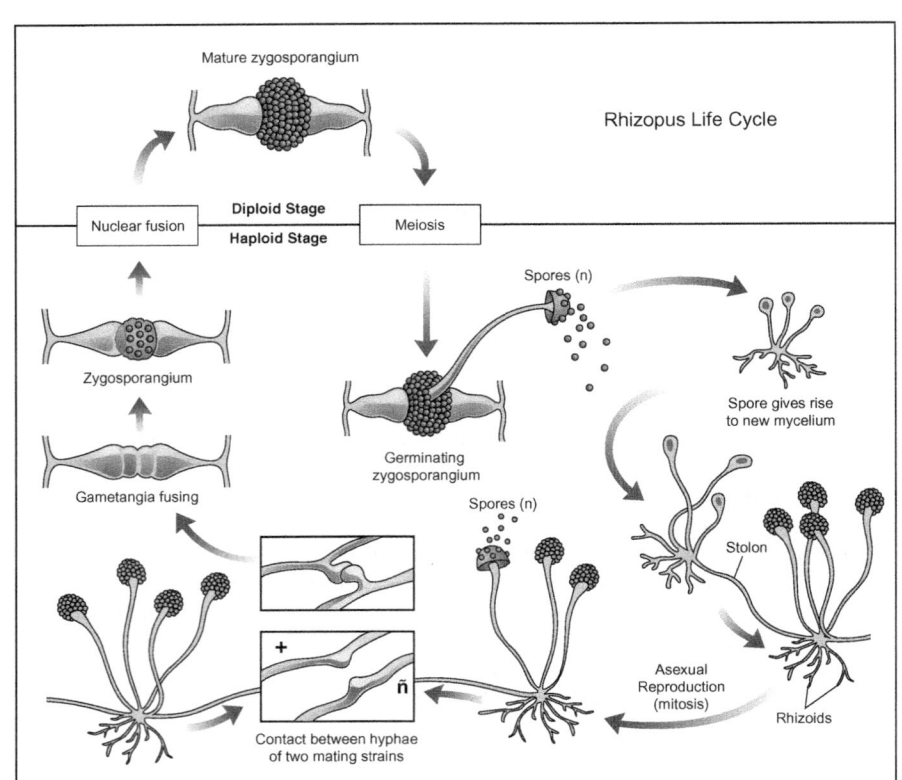

Figure 9. 1 Rhizopus life cycle

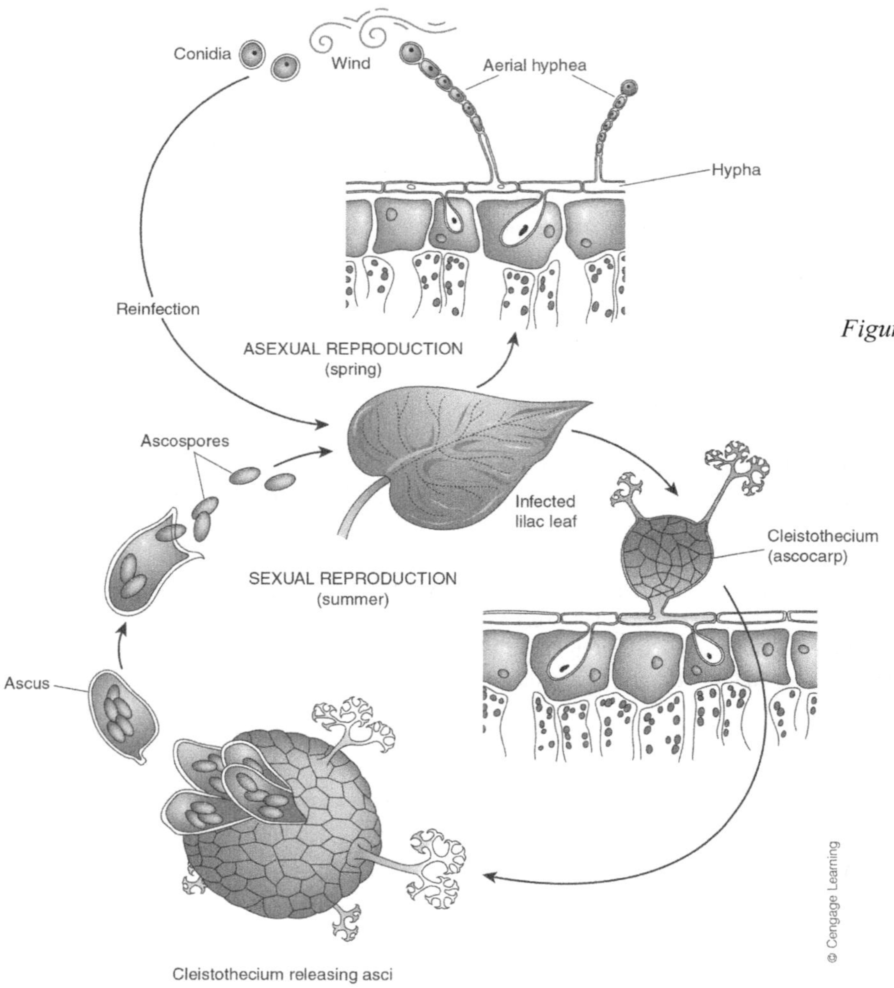

Figure 9. 2 Ascomycota life cycle

9.1 Observation: Zygomycetes

Obtain a Petri dish containing bread mold. View the organisms with your dissecting scope and draw their appearance in the space below. You may see zygosporangia, formed by the fusion of the 2 types (+ and -) of hyphae. The small black, spherical structures are sporangia and form asexual spores.

Now, observe prepared slides of Rhizopus sporangia and zygosporangia and make a sketch below.

POSTLAB QUESTIONS
Q1. In what structure does meiosis occur?

Q2. Considering all of the fungus that you observed on the Petri dish, is most of it haploid or diploid?

9.2 Observation: Ascomycetes

Obtain a yeast suspension from your instructor. Prepare a wet mount of the yeast. View under the microscope. Yeast divides by budding. Ascomycetes can form asexual spores (conidia) as well as sexual spores (ascospores). Draw a cell that is budding and compare it to one that is not. Describe what you see.

Obtain a Petri dish containing a culture of *Sordaria* or other sac fungus. Examine the Petri dish with a dissecting microscope. Note the small, dark, pear-shaped structures. These are called **perithecia** and contain the **asci** and ascospores. Gently remove one of the perithecia and place it on a microscope slide in a drop of water. Examine it under low power of a compound microscope. Then, add a coverslip. Tap on the coverslip with a dissecting needle or the eraser of your pencil to **GENTLY** crush the perithecium and release its contents. Examine the results. Draw what you see in the space below.

Obtain a prepared slide of Peziza, a cup fungus. Make a sketch below and label the asci and ascospores.

PRELAB QUESTIONS

Q1. Describe in your own words the process of budding. Is it sexual or asexual?

Q2. Are many of the cells that you see in your preparation undergoing budding?

Q3. What color are the ascospores?

Q4. How many spores are in each ascus?

Q5. Consider all of the fungus you viewed. Was most of it haploid or diploid?

9.3 Observation: Basidiomycetes

Obtain one of the mushrooms provided in lab. This is the basidiocarp, or fruiting body of the fungus. Mushrooms also have an extensive under-ground mycelium similar to that of bread molds or Sordaria. Sketch the appearance of the basidiocarp, noting the cap, gills, stalk, and annulus (if present). Carefully remove one of the gills and place it on a microscope slide. You should be able to see tiny basidiospores attached to sexual cells called basidia. Draw and label what you see in the space below.

9.4 Observation: Fungi Imperfecti

In this group, you would find examples such as Penicillium and Aspergillus, both of which reproduce asexually by conidia. Obtain a prepared slide of Penicillium and Aspergillus. Make sketches below and label the conidia.

9.5 Observation: Lichen

Lichens are symbiotic relationships between fungi and algae.
There are three growth forms of lichens: **crustose, foliose, and fructicose**
Observe one or two growth forms of lichens. Sketch the different structures of the lichen and add descriptions of the structures you see.

POSTLAB QUESTIONS
Q1. In your opinion, do both the algae and fungi benefit from this relationship? Explain.

Lab 10:
Development in Animals, Part I: Sea Urchins
Carolyn Eberhard

OBJECTIVES
1. List the three important functions of fertilization.
2. Describe the four steps that take place in an invertebrate egg during fertilization and that distinguish an unfertilized egg from a fertilized one.
3. List the four main stages of embryonic development following fertilization.
4. Describe cleavage and describe two important ways in which cleavage in a sea urchin embryo is like that in a human embryo.
5. Correctly explain and use the following words: archenteron (primitive gut), blastocoel, blastopore, blastula, cleavage, gastrula, gastrulation, morula, neurula, neurulation, and zygote.
6. Draw a diagram of an embryo containing the three primary germ layers; label the germ layers, blastocoel, blastopore, and archenteron, and label the embryo with its correct embryological stage.
7. List or recognize the body parts formed from each primary embryonic germ layer.
8. Define metamorphosis and explain the selective advantage to animals of going through metamorphosis at some stage of the life history.

KEY TOPICS

Echinoderms	Induction	Gastrulation
Organogenesis	Cleavage	Extraembryonic membranes
Fertilization	Amniote egg	

INTRODUCTION

You have already studied **meiosis**, the process in which a diploid organism prepares for reproduction by packaging its genetic material into haploid **sperm** and **ova** (eggs). Each **gamete** contains a complete genome, but only half of each parent's genetic material. The process of reproduction is continued by **fertilization**, in which haploid gametes join to make a diploid **zygote**, thereby restoring the original number of chromosomes. Meiosis and fertilization are thus two complementary halves of an animal's life history:

The animal's sex may be determined at the time of fertilization by the combination of chromosomes that results. Fertilization also provides the stimulus for the egg to start its development.

The reproductive cycle is complete when the zygote undergoes development into the many-celled adult organism that becomes sexually mature, undergoes meiosis, and then reproduces in its turn. The orderly process by which a developing organism becomes increasingly complex is known as epigenesis. Although developmental processes continue throughout an organism's life, today's lab will focus on embryonic development, which begins with the formation and fertilization of the egg and ends with a young organism, usually similar in form to the adult. Certain important processes go on in the developing embryo. Cell division by mitosis results in the multitude of cells comprising the adult. Cell differentiation, the expression of a different set of genes in different cell types, provides the spectrum of cell types necessary in the adult. Through cell movement, the cells form a highly organized structure that permits the cells of the adult to function properly. All of these processes mutually interact.

STRUCTURE AND FUNCTION IN THE SEA URCHIN

If live sea urchins are available, your instructor may ask you to complete this section on the sea urchin as a representative echinoderm. Al- though echinoderm adults are not at all like humans, early development in these deuterostome organisms are similar to that in humans in several important ways.

- Observe the sea urchins in the aquarium or supply dish. The spines that cover the top (aboral) surface of the body are mobile, and can pivot on little bumps at the base of each spine. If the urchin is healthy, the spines will stick straight out. If the urchins are crawling on the surface of the aquarium, you will be able to see the tube feet sticking to the glass and the mouth, which is located on the flat underside (see *Figure 10.1*).

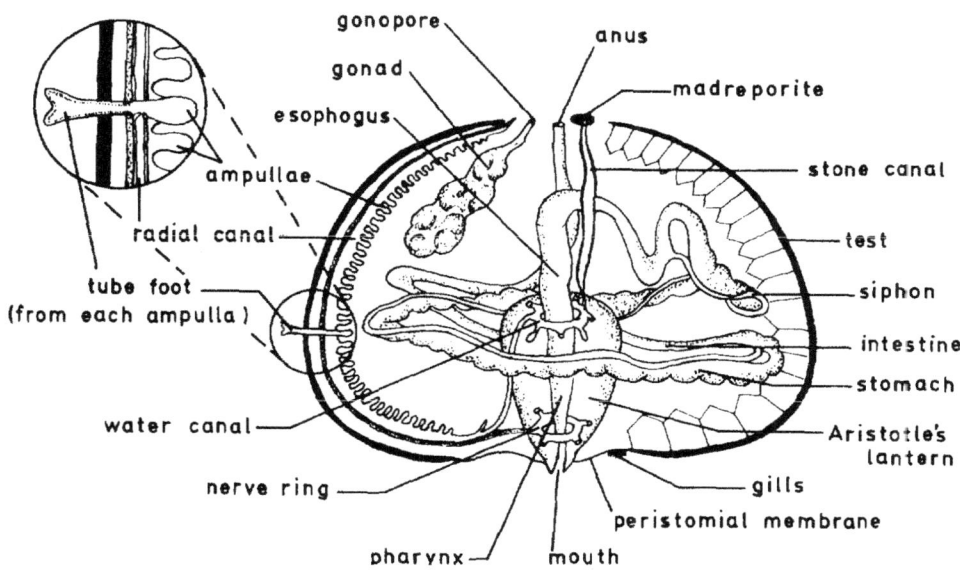

Figure 10.1 A sea urchin.
Aristotle's lantern is the feeding apparatus used for ingesting kelp and seaweed. The gonads, test (skeleton), and water-vascular system show pentaradial or five-fold symmetry, although diagrammed only once. The many movable spines that cover the surface of the sea urchin are not shown.

PRELAB QUESTIONS

Q1. Can you tell which sea urchins are male and which are female?

Q2. Why not?

ANATOMY OF THE SEA URCHIN

*This section will be supplemented with a hand out from your TA

Embryonic development usually ends with the hatching of the embryo or its birth when it leaves the body of its mother. Some organisms, such as the sea urchin, insects, and some amphibians, pass through a developmental stage called a larva, which is different in form from the adult, but usually has its own independent existence. Metamorphosis is the process by which a larva develops into a very different young adult. Humans and sea urchins are both deuterostomes (second mouth) because they share several features of embryological development:

1. The blastopore becomes the anus and the mouth develops later
2. Cleavage of the zygote is radial and indeterminate

10.1 Observation: Sea Urchin Anatomy

*For this section you will be observing a live animal, HANDLE THE URCHINS WITH CARE

Place the urchin in a beaker of seawater or on a Petri dish of sea water with the flat (oral) side down. Observe the animal for a few minutes, and notice the tiny, clear tube feet that extend from between the spines and stick to the surface of the glass.

1) Note how the water-vascular system of radial canals and ampullae connects to all of the tube feet. This is a hydraulic system that allows the feet to be extended by increasing the pressure within the ampulla at the base of each foot. When the foot contacts something, suction can be exerted so that the foot sticks to the object like a tiny suction cup. Each foot can be operated individually. Water circulates into and out of the water-vascular system through a porous plate called the madreporite, which is located on the aboral surface opposite the mouth. Also located on the aboral surface are the anus and the five genital pores through which gametes are released from the gonads into the sea. The shape of the sea urchin is due to a hard endoskeleton called the test. Both the test and the spines are covered by a delicate epidermis. Study the sea urchin test on demonstration, using the dissecting microscope when appropriate.

2) Locate the five gonopores on the aboral surface.

3) Notice the tubercles where the spines were once attached.

4) Find the numerous double perforations where the tube feet extended through the test. These occur in five regions that are separated from each other by five sectors in which there were no feet and thus no perforations.

   By now you should be able to describe the type of symmetry that the sea urchin shows (twofold/threefold/fourfold/fivefold/sixfold) (bilateral/radial):

5) On the oral surface notice the large opening in the test. This is where the feeding apparatus, called Aristotle's lantern, originally protruded through the test, but the membranes holding it together and in position disintegrated when the test was prepared. You can see this structure in your living animal.

6) Again using forceps or gloves, invert the animal so that the flat side is on top. The animal is now upside down from its normal position.

7) Note the feeding structure, Aristotle's lantern, in the center of the oral surface. It contains five teeth that are used to tear the kelp on which the animal feeds, and it is held in place by a membrane that attaches it to the skeleton.

8) When you are finished observing the structure of the oral surface, leave your animal upside down for awhile, and observe its response from time to time.

POSTLAB QUESTIONS

Q1. How did your sea urchin react to an upside down situation?

Q2. What structures were used in the response?

When you are finished with your sea urchin, leave the animal upside down and prepare your station for the *in vitro* fertilization experiment.

SEA URCHIN DEVELOPMENT

Mature sea urchins are male or female. When the gametes are ripe, they are released through pores on the top (aboral) side of the animal, and fertilization takes place in seawater.

In the laboratory, the urchins can be induced to release their gametes separately so that fertilization can be carried out and observed under the microscope. If the fertilized eggs are carefully washed and incubated in seawater, they will develop for many days, and will eventually form a larval stage known as the pluteus. The pluteus is a free-swimming, feeding organism that eventually attaches to the bottom and metamorphoses into a tiny sea urchin. In this organism the main function of the larval stage is dispersal; the sessile (non-motile) adults accomplish most of the feeding and movement. Depending on the time of year, you will use an East Coast sea urchin (Arbacia) or a West Coast urchin (Strongylocentrotus) to study early development. The schedule for the development of these species is shown in *Table 10.1*.

Cleavage begins with fertilization and continues through many cell divisions without growth until the embryo is a hollow ball of many cells. Sea urchin eggs are very large in size because they contain everything needed for cleavage: ribosomes, mRNA information for the proteins of early development, and energy stored in the form of yolk granules. Because cleavage is so independent of outside needs, it can be induced artificially without any sperm, a phenomenon known as parthenogenesis. The genetic contribution of the sperm becomes important only later in development. You should use a depression slide or omit the cover slip while you are observing the eggs and embryos, because they are so large. **Note: If you do not use a cover slip, your sample runs the risk of drying out, as the water will evaporate. This will cause salt crystals to form and your embryos to die, so make sure your sample stays wet by rehydrating the slide with drops of salt water.**

The eggs are chemically attractive to sperm and are soon surrounded by swarms of these relatively tiny cells. When the sperm cells penetrate the jelly coat, they become activated by the acrosome reaction so they can enzymatically digest their way through the jelly coat. Eventually they come into contact with the membrane of the cell and cell recognition occurs. Then one lucky sperm head will be drawn into the cytoplasm by the fertilization cone that forms, leaving its tail behind. Its nucleus will combine with the egg's nucleus to form the new zygote nucleus. The egg and sperm contribute an equal number of chromosomes to the zygote. Finally, activation of the egg starts the developmental process. **You can barely see the individual sperm cells under low power, but you can see right away that something is happening.**

Table 10.1 Development of Sea Urchin Embryos

Developmental Stage	*Arbacia punctulata* at 23°C	*Strongylocentrotus purpuratus* at 15°C
Fertilization	0	0
Fertilization membrane forms	1–2 min	1–2 min
First cleavage	50–70 min	90–110 min
Second cleavage	80–110 min	140–170 min
Blastula	5–6 hr	8–10 hr
Hatching of blastula	7–8 hr	18–20 hr
Gastrula	12–15 hr	22–26 hr
Pluteus	24 hr	72 hr
Metamorphosis	2 weeks	2–8 weeks

When the egg cell is stimulated by the arrival of sperm, a chain of events begins to unfold that is very precise and almost invariably ends in the production of a normal sea urchin pluteus. First an electrical reaction begins when Na^+ channels open at the site where the sperm penetrates, and it spreads in the cell membrane all over the surface of the cell. Changes in the cell membrane prevent any second or third place sperm heads from entering the cytoplasm. Then Ca^{++} ions released from this region of the membrane induce a cortical reaction in the egg cytoplasm, signaling that development should begin. The vitelline membrane, which lies between the cell membrane and the jelly coat, separates from the cell membrane and enlarges so that there is a wide space between the two membranes. The cell "membrane" now looks double and is called the fertilization membrane. This is the obvious change that you see whenever fertilization occurs. Any sperm still struggling through the jelly coat are lifted away from their goal and the fertilization membrane provides a protected environment for the developing embryo. The egg and sperm pronuclei fuse to form the nucleus of the zygote. Rapid protein synthesis is carried out using mRNA provided by the mother and stored in the egg. This is part of the preparation necessary for the first cell division, when the constriction known as the cleavage furrow will divide the zygote into two cells.

Once the first division has taken place, cleavage will continue for many cell divisions. In sea urchins, cleavage is indeterminate: each of the cells formed early in development has the potential to develop into a complete organism. If the daughter cells should happen to separate after the first division, the result would be two identical organisms. When this happens in human development, the result is identical twins. Cleavage is radial because each cleavage plane is at right angles to or parallel to the original axis of the zygote. In the sea urchin the yolk is in the form of cytoplasm granules, so the cleavage furrow can pass completely through the egg, from the animal pole (top) to the vegetal pole (bottom), without difficulty. This type of cleavage is holoblastic, which has three layers. In this process, cell movement is extremely important. Cells migrate into the interior of the blastula at a point called the blastopore, and form a lining so that the embryo now has two germ layers: an outer ectoderm and an inner endoderm. The ectodermal cells will form the epidermis or skin; they will also form the neurectoderm, which will give rise to the whole nervous system, including the brain. The endoderm develops into the lin- ing of the gut and of certain digestive organs and glands. The space inside both layers is called the archenteron, and will eventually form the lumen of the organism's gut. The cells in the dorsal part of the endoderm form pouch, and bud off to become the cells of the mesoderm. The interior of the pouches forms the coelom by enterocoely. This type of mesoderm formation is an important way (besides radial, indeterminate cleavage, and formation of the anus from the blastopore) in which sea urchin embryology and human embryology are alike. Mesoderm gives rise to the skeleton, muscles, sex organs, and other internal organs of the adult.

The morula is a solid ball of cells that gradually develops a hollow interior called the blastocoel. The embryo is then at the blastula stage, and its outer cells are ciliated so that the blastula is motile, and can move about within the fertilization membrane. (The human equivalent of the blastula is not motile and is called the blastocyst. It moves along the oviduct by the action of the mother's cilia and implants in the wall of the uterus about 7 days after fertilization.) After several hours of cleavage, the sea urchin blastula is fully developed, and "hatches" as the fertilization membrane finally breaks open and releases the motile embryo. Cleavage has ended and the blastula now enters the next stage of embryonic development.

Stages of Development

Gastrulatiom
During gastrulation the embryo changes from a blastula with one layer of cells into a gastrula,

Neurulation
In vertebrates such as yourself and the chick, the gastrula soon develops a dorsal groove, which pinches off and becomes part of the dorsal nerve cord and the brain. An embryo undergoing this process is at the neurula stage.

Organogenesis
If older sea urchin embryos are available, you will be able to see a simple digestive system and skeletal system.

Metamorphosis
The sea urchin embryo develops skeletal elements, and begins feeding as the pluteus larva. It continues this existence for up to four months until it finally settles down on the substrate, attaches, and changes into a tiny adult sea urchin. The whole process of metamorphosis may take only an hour, and the result is a miniature adult only 1 mm in diameter

POSTLAB QUESTIONS

Q1. What type of fertilization do sea urchins use?

Q2. What change occurs within 1 or 2 min of adding sperm cells to the eggs?

Q3. Are the sperm cells in motion? Can you see the flagella of the sperm cells?

Q4. In addition to the nucleus and the flagellum, what organelle is important for sperm cells to carry out their reproductive role? (Hint: This organelle provides energy needed in large amounts for sperm motility.)

Q5. How large is the embryo compared to the egg after several divisions? Can you see the primitive gut? How did the embryo obtain energy before its gut formed to permit feeding?

Q6. What is the advantage of a motile larval stage in this species?

Lab 11:
Development in Animals, Part II: Chick Development
Carolyn Eberhard

This entry in the lab manual will serve as background material for your experiment. The purpose of this lab is to view development of the chick embryo; therefore, the majority of your time should be spent noticing differences between the different stages of development.

FERTILIZATION AND EARLY DEVELOPMENT

The ovum (egg cell) of a chicken consists of a yellow mass of yolk and a small area of yolk-free cytoplasm, the blastodisc, that will develop into the embryo proper. Cleavage and development occur only within the disc, and the yolk supplies required nutrients and energy to sup- port the embryo. As the egg cell passes through the reproductive tract of the female chicken, it is modified in several ways. A solution of the protein albumin is added to make up the white of the egg. This layer provides physical protection for the delicate embryo, and antimicrobial sub- stances in it protect the embryo from bacterial infection. The albumin itself will provide the chick with a store of amino acids to draw on for protein synthesis, and the liquid is the chick's only source of water. Later on, two membranes and then a porous shell are added. The hard shell protects the embryo and its food supply but also allows gas exchange with the environment. The egg of a bird is an amniote egg. Unlike a fish or amphibian egg, it is fully adapted to terrestrial reproduction. Reptiles and mammals also produce amniote eggs in which the embryo is protected in a miniature watery environment within the amnion membrane.

If the egg cell has been fertilized, it will begin cleavage and start to develop right in the hen's oviduct. The cleavage furrows will not be able to pass through the massive yolk, however, restricting cell division to the blastodisc. This type of cleavage is meroblastic. Since eggs are held in the oviduct for various lengths of time before being laid, the exact stage of develop- ment at the time of laying is different for each individual egg, but usually cleavage, blastula formation, and gastrulation have already been accomplished before the egg is laid.

Early in development, the cells of the blastodisc split into two layers, like a pita bread: the upper layer is the epiblast and will become the ectoderrn, and the lower layer is the hypoblast and will become endoderm. The blastodisc now has a blastocoel, but it is flat rather than a hollow ball, like the sea urchin blastula. A thickening called the primitive streak then develops in the center of the blastodisc. Here cells migrate from the epiblast into the space of the blastocoel and form the embryo's mesoderm.

Mesodermal tissue near the center of the embryo pinches off to form the notochord and the pairs of somites that lie in rows along each side of the notochord. The notochord induces the ectoderm above it to form a neural plate. The plate develops folds, which eventually meet and fuse to form the neural tube. The neural tube in turn differentiates into the brain and the spinal cord. Finally the embryo lifts, and the ectoderm folds beneath it so that the embryo is completely covered by ectoderm and the primitive gut is enclosed by the endoderm. These changes are shown in *Figure 11.1*.

The notochord of the gastrula induces part of the ectoderm to invaginate and form the neural tube. The brain and nervous system will eventually differentiate from the neural tube; meanwhile, specialized cells in the area of the blastoderm surrounding the embryo form blood islands, which give rise to extraembryonic blood vessels and blood cells. The embryo develops a primitive heart, and blood soon begins to circulate between the embryo and the surface of the yolk, where capillaries allow gas exchange and absorption of nutrients.

Other cells in the blastodisc grow into protective extraembryonic membranes. Although these membranes are not part of the embryo proper, they are vital for its nourishment and development.

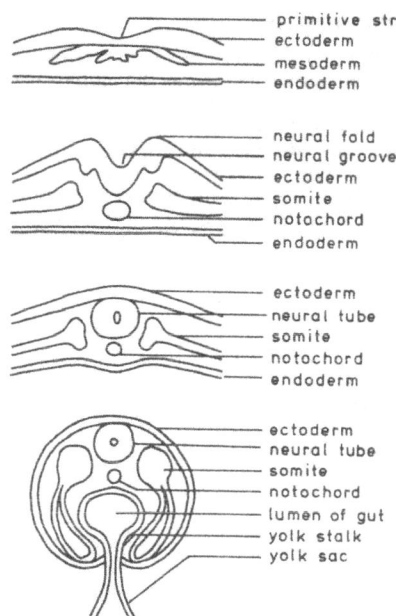

Figure 11.1 Chick embryo in cross section.

The notochord of the gastrula induces part of the ectoderm to invaginate and form the neural tube. The brain and nervous system will eventually differentiate from the neural tube.

You can color-code structures derived from the primary germ layers: ectoderm = blue, mesoderm = red, and endoderm = yellow.

Fertilization and Early Development

1) The amnion develops as the inner layer of a fold of tissue growing down over the embryo's head. It meets with another fold growing up over the tail, and fuses with it so that the embryo is completely enclosed in an amniotic sac, which is filled with amniotic fluid. The fluid acts as a protective cushion. The growing amnion is visible as a fold of tissue in the 48-hr chick.

2) The yolk sac grows out from the primitive gut, and eventually surrounds the whole yolk with a double membrane. Blood vessels in its walls are important for the nutrition of the growing embryo. The blood vessels you will see in the 48- and 72-hr chicks develop in the yolk sac and are important in respiration as well as nutrition at this time.

3) The chorion is the outer layer of the fold of tissue that forms the amnion and can be seen at 48 hr. It will eventually surround the amnion, embryo, and yolk; it protects the embryo and holds it in contact with the yolk. Later it fuses with the allantois to form the chorioallantoic membrane, the respiratory organ of the chick during later development. This membrane completely covers the embryo and yolk by day 12 of development.

4) Finally the allantois grows out as a sac-like extension of the primitive gut, expands up over the embryo, and eventually surrounds the embryo and yolk with a double membrane. The outer layer of the allantois presses up against the chorion, and the inner layer against the amnion and yolk sac. Because the allantois is richly supplied with blood vessels, it allows gas exchange with the external environment. The wastes that accumulate during development, mostly in the form of uric acid, are stored in the interior of the allantois until the chick hatches. The allantois can be seen as a small sac in the 72- and 96-hr chick embryo.

Important cellular processes underlie these visible changes during development. Cells must divide, stop dividing, and even die at the appropriate time. They must move about, recognize, and adhere to their neighbors properly. Differentiation of cells depends on their patterns of movement and position. It is often brought about by embryonic induction in which one cell alters the fate of another through complex chemical and surface interactions.

Live Chick Embryos

Each pair of students will be provided with chick embryos that have been incubated for 33, 48, 72, and 96hrs. You are responsible for understanding all of these stages, so study the prepared samples or another student's embryo to see all the structures. You may also be presented with older chicks (I.E. Developed for 120hrs).

1.) Add a little warmed saline (0.31% NaCl) to a finger bowl to a depth of a centimeter or so. It should not cover the yolk of the egg you are going to open in it. (Saline may be omitted if you work quickly.)
2.) Take an egg from the incubator or tray, hold it in exactly the same orientation, and carry it carefully to your desk.
3.) Turn the egg sideways and wait 1 min for the yolk to adjust.
4.) Do not tap the egg on the side of the dish, as this will cause jarring of the embryo. Instead, use a needle and poke the shell gently.
5.) Use tweezers to gently remove the shell and release the contents into the dish without disturbing the yolk..

If your embryo is 72 hr old, you should see the embryo and its beating heart immediately. If you do not see your embryo, ask your instructor for help. Either the egg failed to develop, or the yolk is turned upside down so that the embryo is hidden.

Important Features in Chick Development

33-Hour Chick (*Figure 11.2*)

IMPORTANT FEATURES to look for in the 33-hr chick embryo include 12 somites, heart, vitelline veins, notochord, neural tube, optic vesicles, and blood islands. The embryo is a streak in the center of the blastodisc, and you will have to transfer it to a watch glass to see much.

The notochord has formed in the mesoderm and has already caused the neural tube to form in the neurectoderm by embryonic induction. This process begins and is more advanced in the head region. The brain has begun to differentiate, and the optic vesicles are developing. There are about 12 somites along the sides of the notochord, which induced them to form from mesoderm. They will form muscle and parts of the skeleton.

Because the tail region develops more slowly than the head regions, it may still show the primitive streak that was the very first sign of development. The vitelline veins have joined to form the primitive heart. It might be feebly contracting, but there is no blood flow yet. If you are looking at a stained slide, notice the blood islands in the blastodisc outside of the embryo itself. They will give rise to the blood cells and some of the vessels. The amnion and chorion are beginning to develop as a fold of tissue growing down over the head region. It will eventually contact the posterior fold to enclose the embryo. Finally you may be able to see the foregut, which is growing toward the head region from underneath the embryo as an outpocketing of the endoderm. It will later extend toward the tail region as well. The anterior end of the neural tube has begun to differentiate while the posterior end has not yet formed from the neural groove.

Examine the demonstration of a vertebra to be sure that you understand the relation- ship of the notochord and the spinal cord. The notochord gives rise to intervertebral discs, and the spinal cord of the neural tube runs through the opening in the vertebra.

Q1. From which germ layer did the spinal cord develop?(ectoderm/mesoderm/endoderm)

Q2. What structures will give rise to the vertebrae and back muscles?

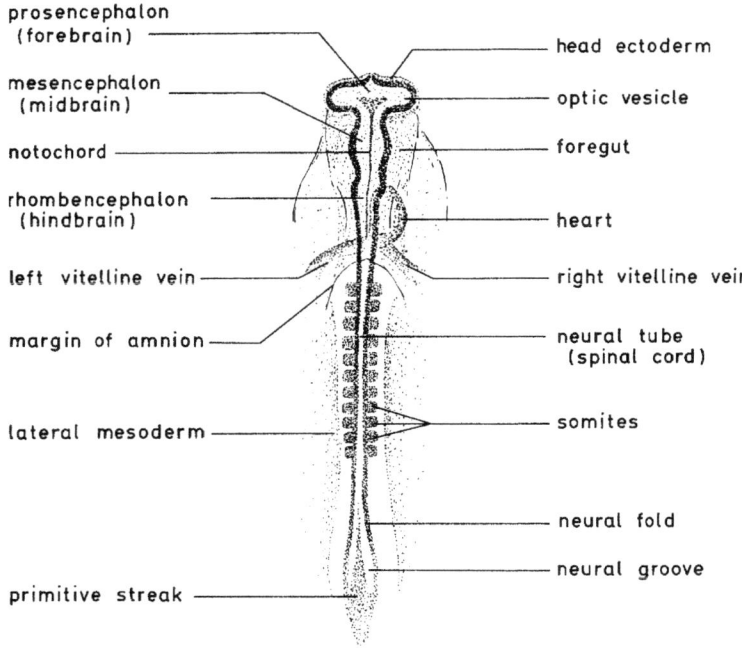

Figure 11.2 Whole mount of a 33-hour chick embryo.
The anterior end of the neural tube has begun to differentiate while the posterior end has not yet formed from the neural groove.

48-Hour Chick (*Figure 11.3*)

IMPORTANT FEATURES to look for in the 48-hr chick embryo include 28 somites, eye with lens, distinct telencephalon and diencephalon, aortic arches, vitelline arteries, amnion, and chorion. This embryo has started to twist to the right so that the head region is now lying on the chick's left side. This is the process of torsion, which will continue until the whole embryo is lying on its left side. At the same time, the embryo is curling up in the process of flexure so that the neural tube takes the shape of a question mark with a bend in the region of the hind-brain. The forebrain has formed the diencephalon and telencephalon. Blood circulation usually begins at about the 16-somite stage.

If you are viewing a living embryo, look for contractions of the heart and circulating blood.
Trace the flow of blood from the ventricle of the heart through the aorta, and then through the aortic arches in the region below the head (pharynx).

Identify the parts of the brain: The optic cup has caused embryonic induction of the lens of the eye by the ectoderm, and the neural tube has induced the formation of the otocyst, or primitive ear. In each case a chemical signal called an inducer caused the overlying ectoderm to differentiate into an appropriate structure.
Note the vitelline arteries growing out over the yolk. They join the vitelline veins in capillary beds on the yolk surface. The margin of amnion and chorion can be seen above the vitelline arteries.

Q1. How many somites are still showing their dorsal sides?

Q2. What structure induced the somites?

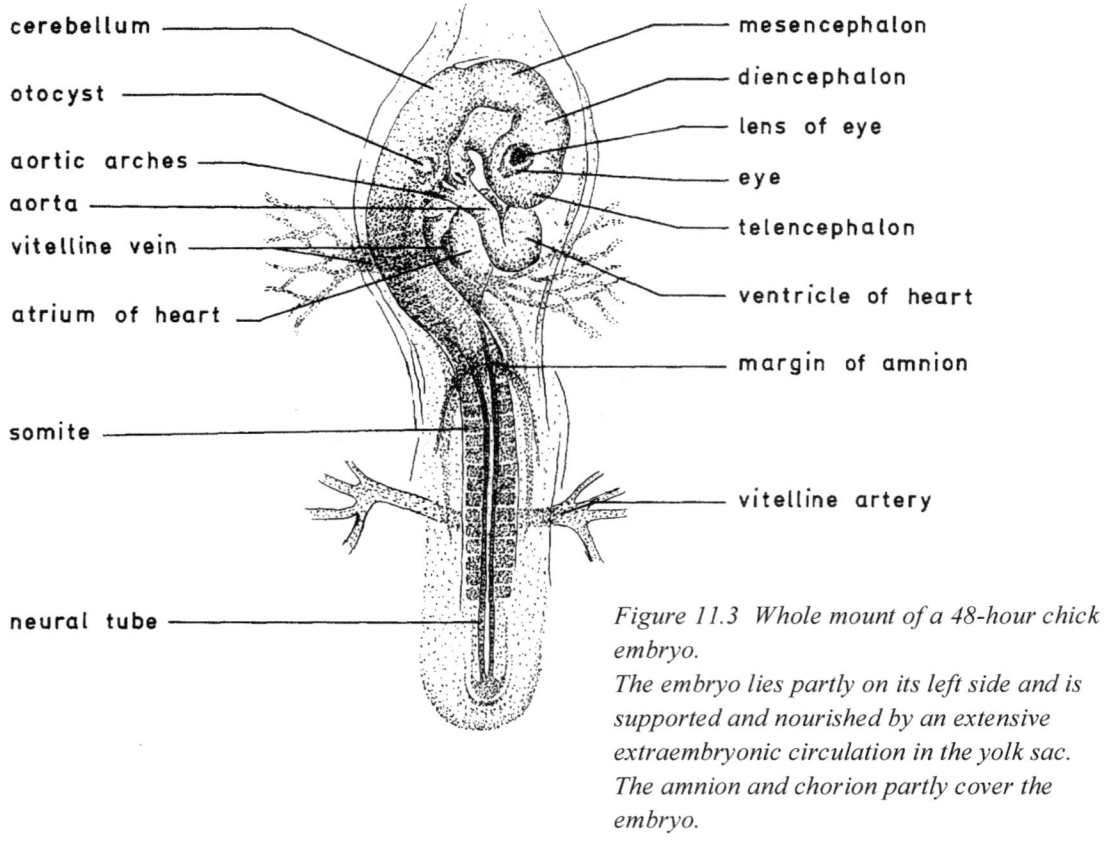

Figure 11.3 Whole mount of a 48-hour chick embryo.
The embryo lies partly on its left side and is supported and nourished by an extensive extraembryonic circulation in the yolk sac. The amnion and chorion partly cover the embryo.

72-Hour Chick (*Figure 11.4*)

IMPORTANT FEATURES to look for in the 72-hr chick embryo include 36 somites, extraembryonic circulation in yolk sac, mesencephalon, cerebellum, otocyst, pharyngeal clefts, and limb and tail buds.
The chick is now almost completely on its left side (torsion). Flexure of the brain is prominent, and it is differentiated into distinct regions. There are too many somites to count easily (about 36).

Note the optic cup, the lens, and the otocyst, or primitive ear. The atrium and ventricle of the heart are greatly enlarged and are actively pumping blood through the vitelline and embryonic arteries.

Below the heart you can see the anterior limb buds, which will develop into wings, and near the embryo's tail you see the posterior limb buds, which will form the legs. You may be able to see the small sac-like allantois in the region of the posterior limb buds, but it is probably hidden by the tail, and is better seen in later embryos. If you can successfully transfer your embryo to a watch glass, observe circulation in the blood vessels under low power with your microscope.

Q1. What membranes completely enclose the embryo itself?

Q2. Where does the circulating blood pick up oxygen and nutrients?

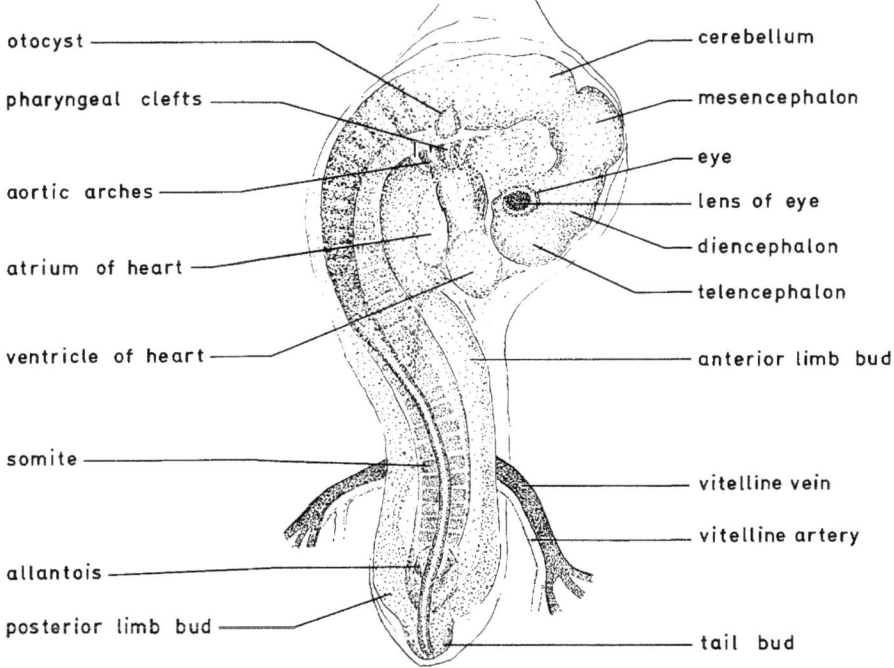

*Figure 11.4 Whole mount of a 72-hour chick embryo.
The embryo lies mostly on its left side and the allantois and limb buds are beginning to develop.*

96-Hour Chick

IMPORTANT FEATURES of the 96-hr chick embryo include 41 somites, allantois, and limb and tail buds. An embryo at this stage has de- veloped about 41 somites and the yolk is partly covered with prominent blood vessels.

The flexure of the body posterior to the brain is now obvious. The eye is easier to see because it is pigmented, but the ear is less distinct because the tissues have become much thicker.

Remove the membranes surrounding the embryo so that you can see the heart and limb buds. You can use a probe or teasing needle to straighten out the embryo so that you can see the pharyngeal clefts and pouches. The fluid-filled sac at the posterior end is the allantois. It will expand and eventually fuse with the chorion to become the highly vascularized chorioallantoic membrane. This membrane remains close to the porous shell of the egg. When the chick hatches, the chorioallantois will remain behind as a waste-filled membrane stuck to the shell, but its base within the embryo gives rise to the urinary bladder.

Q1. What are the two functions of the allantois?

Q2. Why does the chick secrete nitrogenous wastes as uric acid whereas humans secrete mostly urea?

Later Development

Your instructor may demonstrate some older living embryos that are even further along in development. All of the major developmental processes happen during the first 6 days of development. The embryo then grows and matures during the rest of the incubation period until it hatches at 21 days and begins its independent existence as a wobbly, fluffy chick.

When you are finished with your work, discard your debris in the EGG WASTES container, clean your dissecting instruments, return the prepared slides, and store your microscopes properly.

POSTLAB QUESTIONS

Q1. Is fertilization in the chicken internal or external?

Q2. Is there any nutritional advantage in eating fertilized eggs?